Energiepolitik und Klimaschutz
Energy Policy and Climate Protection

Reihe herausgegeben von
L. Mez, Berlin, Deutschland
A. Brunnengräber, Berlin, Deutschland

Diese Buchreihe beschäftigt sich mit den globalen Verteilungskämpfen um knappe Energieressourcen, mit dem Klimawandel und seinen Auswirkungen sowie mit den globalen, nationalen, regionalen und lokalen Herausforderungen der umkämpften Energiewende. Die Beiträge der Reihe zielen auf eine nachhaltige Energie- und Klimapolitik sowie die wirtschaftlichen Interessen, Machtverhältnisse und Pfadabhängigkeiten, die sich dabei als hohe Hindernisse erweisen. Weitere Themen sind die internationale und europäische Liberalisierung der Energiemärkte, die Klimapolitik der Vereinten Nationen (UN), Anpassungsmaßnahmen an den Klimawandel in den Entwicklungs-, Schwellen- und Industrieländern, Strategien zur Dekarbonisierung sowie der Ausstieg aus der Kernenergie und der Umgang mit den nuklearen Hinterlassenschaften.
Die Reihe bietet ein Forum für empirisch angeleitete, quantitative und international vergleichende Arbeiten, für Untersuchungen von grenzüberschreitenden Transformations-, Mehrebenen- und Governance-Prozessen oder von nationalen „best practice"-Beispielen. Ebenso ist sie offen für theoriegeleitete, qualitative Untersuchungen, die sich mit den grundlegenden Fragen des gesellschaftlichen Wandels in der Energiepolitik, bei der Energiewende und beim Klimaschutz beschäftigen.

This book series focuses on global distribution struggles over scarce energy resources, climate change and its impacts, and the global, national, regional and local challenges associated with contested energy transitions. The contributions to the series explore the opportunities to create sustainable energy and climate policies against the backdrop of the obstacles created by strong economic interests, power relations and path dependencies. The series addresses such matters as the international and European liberalization of energy sectors; sustainability and international climate change policy; climate change adaptation measures in the developing, emerging and industrialized countries; strategies toward decarbonization; the problems of nuclear energy and the nuclear legacy.
The series includes theory-led, empirically guided, quantitative and qualitative international comparative work, investigations of cross-border transformations, governance and multi-level processes, and national "best practice"-examples. The goal of the series is to better understand societal-ecological transformations for low carbon energy systems, energy transitions and climate protection.

Reihe herausgegeben von

PD Dr. Lutz Mez
Freie Universität Berlin

PD Dr. Achim Brunnengräber
Freie Universität Berlin

Weitere Bände in der Reihe http://www.springer.com/series/12516

Sevinj Amirova-Mammadova

Pipeline Politics and Natural Gas Supply from Azerbaijan to Europe

Challenges and Perspectives

 Springer VS

Sevinj Amirova-Mammadova
Baku, Azerbaijan

Dissertation Freie Universität Berlin, Department of Political and Social Sciences,
Otto Suhr Institute of Political Science, 2017

D 188

Energiepolitik und Klimaschutz. Energy Policy and Climate Protection
ISBN 978-3-658-21005-2   ISBN 978-3-658-21006-9   (eBook)
https://doi.org/10.1007/978-3-658-21006-9

Library of Congress Control Number: 2018932744

Springer VS
© Springer Fachmedien Wiesbaden GmbH, part of Springer Nature 2018

Printed on acid-free paper

This Springer VS imprint is published by the registered company Springer Fachmedien Wiesbaden
GmbH part of Springer Nature
The registered company address is: Abraham-Lincoln-Str. 46, 65189 Wiesbaden, Germany

Acknowledgments

There are several people to whom I would like to express my gratitude for their contribution and support. First and foremost, my thanks go to my supervisor, PD. Dr. Lutz Mez, who provided me with constant encouragement and sage advice from the very beginning of this project. His invaluable suggestions also helped me to deal with a ruthless critique of all the chapters provided by my colleagues and friends. I should also thank Tina Flegel, my friend and colleague for her encouragement and support.

Despite a very stimulating academic environment the main challenge consisted in finally finishing the writing of this thesis. My family and friends assisted me enormously in overcoming this challenge. I am extremely grateful for their moral support and practical advices, which greatly smoothed the progress of this at times uneasy process. The debt that I owe to my family is immeasurable.

I withhold my deepest gratitude for my parents, Rafael and Ummahani, my siblings, Zaur and Sevda, and my husband, Mukhtar. Without their support and patience this work would have been impossible. They have always provided me with the unconditional – emotional and practical – support in all my undertakings. I owe my thesis to my family.

Table of Contents

List of Abbreviations

AGRI	Azerbaijan – Georgia – Romania Interconnector
AIOC	Azerbaijan International Operating Company
bcm	Billion cubic meters
BP	British Petroleum
BTC	Baku Tbilisi Ceyhan Pipeline
BTE	Baku Tbilisi Erzurum pipeline
CCGT	Combined Cycle Gas Turbines
CIS	Commonwealth of Independent States
CNG	Compressed Natural Gas
CPC	Caspian Pipeline Consortium
DESFA	Greek Public Gas Transmission System Operator
EC	European Commission
EU	European Union
FDI	Foreign Direct Investment
FSU	Former Soviet Union
GUEU	Georgia-Ukraine-European Union
IAP	Ionian Adriatic Pipeline
IEA	International Energy Agency
IGA	Intergovernmental Agreement
IGB	Interconnector Greece Bulgaria
INOGATE	Interstate Oil and Gas Transport to Europe
ITGI	Interconnector Turkey–Greece–Italy
LNG	Liquefied Natural Gas
MOU	Memorandum of Understanding
MOUC	Memorandum of Understanding and Cooperation
NICO	Naftiran Intertrade Company
NIOC	National Iranian Oil Company
NRAs	National regulatory authorities
PSA	Production Sharing Agreement

SCP	South Caucasus Pipeline
SD	Shah Deniz
SD I	Shah Deniz Phase I
SD II	Shah Deniz Phase II
SEE	Southeastern Europe
SEEP	South East Europe Pipeline
SGC	Southern Gas Corridor
SOCAR	State Oil Company of Azerbaijan Republic
TANAP	Trans-Anatolia Pipeline
TAP	Trans-Adriatic Pipeline
Tcm	Trillion cubic meters
TCP	Trans-Caspian Pipeline
TEP	Third Energy Package
TPA	Third Party Access
TPAO	Turkish Petroleum Corporation
TRACECA	Transport Corridor Europe Caucasus Central Asia
TSOs	Transmission system operators
USA	United States of America

Figures and Maps

1 Introduction

1.1 Background: Rise of natural gas industry and formation of regional markets

The world's energy landscape is continuously evolving as a result of advancing technologies, amplification of environmental concerns, and political and economic crises taking place in different parts of the world. Furthermore, for the last two decades the oil prices have been volatile. The high level of the price instability has been influenced by different factors, including demand and supply dynamics. Global energy demand has started to grow as the need for the transportation fuels, heating sources, and fuels for power generation increased. One of the key factors affecting the growth of the energy demand included economic growth in the developing countries and fuel switching. All these changes forced the search for alternatives.

The major energy development of the 21st century is associated with the rise of the natural gas production and consumption, driven primarily by the industrialization of the economy and the growing global energy demand. The share of the natural gas within the global primary energy mix increased with the discovery of huge reserves at the end of 1990s and advanced Combined Cycle Gas Turbines (CCGT) technologies, which allowed using natural gas as a main fuel or feedstock for power generation. This development in the power generation sector has become a main driver of the growth of the natural gas demand. In comparison to oil and coal, natural gas is easy to handle and cleaner to burn. Low-carbon emissions and low capital cost of natural gas generation compared to other fossil fuel generations, combined with abundant gas supplies and relatively low prices, make natural gas an attractive option in a carbon-constrained environment (MITEI, 2011).

Today, natural gas plays a role of primary fuel for electricity generation by providing base load power and flexible backup to intermittent energy supplies from solar and wind power generations. Consequently, it is considered as a transition fuel ensuring the entrance of the renewals into the global energy mix.

© Springer Fachmedien Wiesbaden GmbH, part of Springer Nature 2018
S. Amirova-Mammadova, *Pipeline Politics and Natural Gas Supply from Azerbaijan to Europe*, Energiepolitik und Klimaschutz. Energy Policy and Climate Protection, https://doi.org/10.1007/978-3-658-21006-9_1

Besides, this the Liquefied Natural Gas (LNG) shipment via tankers over long distances, and the formation of spot markets open new horizons for the development of the global gas market.

Despite its advantages, natural gas industry was often overlooked and underdeveloped in comparison to oil (Stevens, 2010). This was explained by the various factors. First, there was a perception indicating that there were not enough natural gas reserves in the world. Second, natural gas reserves have been found mostly in remote areas and the physical access to the source was often limited. Third, transportation of natural gas from remote areas was technically and financially problematic; exporting natural gas from the field to the market required construction of expensive infrastructure; and for many years pipeline was the only key mode of transportation. Private energy firms often were not interested in natural gas production and delivery, because pipeline construction was very expensive and natural gas was considered as less rent-bringing compare to oil. Therefore, states and state-owned companies were the main actors developing and regulating natural gas industry.

Even today, the development of the natural gas industry and the market are constrained by geographical, political and economic factors. Geographical fragmentation of the natural gas reserves, its regional usage, limited links between supplier and consumer countries, and transportation of the gas over long distances through the pipelines have resulted in regionally segmented markets with different levels of production and consumption. Three regional markets – North America, European including North Africa and Commonwealth of Independent States (CIS), and Asia with the link to the Persian Gulf – can be distinguished (MITEI, 2011). Each of these regions has developed specific pricing models and market structures with the level of market maturity and import dependency determined by geopolitics. The existence of different market structures and their weak integration with each other prevented the formation of a global gas market. That means, the situation in the world's gas market was quite different compare to the global oil market.

State involvement in the natural gas trade, limited transportation options of its export and immaturity of gas markets make natural gas the subject of political interests of certain actors, which tend to use it as leverage. On the other hand, it may result in vulnerability and dependency of consumers on suppliers. There is a need to state that levels of vulnerability and import dependency among suppliers and consumers are different in each of these big regional markets. In contrast to European and Asian markets the North American natural gas market is more complex and developed. Short-term contracts indexed to the primary spot market prices become the primary model of pricing that makes the gas trade more transparent and flexible. European and Asian gas markets are relatively young compare to American market and both of these markets have specific features and frequently face supply shortages.

The Asian gas market is extremely important for LNG, where the share of pipeline gas constitutes almost 20 percent of the imported gas. LNG prices are set through oil-indexed long term-contracts and remain bound to this market structure. The lack of indigenous natural gas resources increases the region's dependency on imported LNG from Southeast Asia, Australia and the Persian Gulf. This dependency places a high premium on security of supply, which is reflected in the region's dependence on long-term, relatively high-priced contracts indexed to oil (MITEI, 2011). But yet political interests are not strongly involved in the Asian gas trade. The key problems are the lack of spot markets and oil indexed prices. Indeed, oil prices' volatility makes oil imperfect index for natural gas[1].

The European gas market that will be more broadly discussed in this study totally differs from the Asian and American markets. It represents a model of a gas market, which is less flexible and more vulnerable to political instability because of the prevalence of pipeline transportation. Three key factors can be determined leading to the vulnerability of the market. First, natural gas supply sources are very limited. Europe imports about 80 percent of its natural gas from Russia, Algeria and Norway. Second, pipeline transportation constitutes the primary share in gas export. About 60 percent of natural gas is transported to European consumers by pipelines traversing territories of transit countries (BP, 2014). Despite developing LNG supplies, cross-border pipelines represent the primary means of transportation of natural gas to Central and South Eastern Europe. Finally, natural gas prices have been agreed based on long-term contracts indexed to the price of crude oil, as in the case of the Asian gas mar-

---

[1]   Since the spot market oil and natural gas price relationship does not match any simple formula, an oil-indexed contract price cannot mimic very well the spot natural gas price; oil indexed prices are out of sync with the value of marginal deliveries of natural gas, sometimes being too high and other times too low. Therefore they cannot give the right signals for consumption of natural gas, inhibiting efficient use of the resource. In order for both buyers and sellers to capture the full value of natural gas resources, it is essential for long-term contracts to reflect the specific supply and demand conditions of natural gas, meaning a liquid market in gas spot deliveries. MITEI. (2011). *The Future of Natural Gas - An Interdisciplinary MIT Study.* MIT Energy Initiative.

ket. Relative inflexibility of the European market, supply dependency, and long supply chains increase the significance of source diversification and impose enhancing energy supply security. In order to reduce energy vulnerability, the EU has to expand its energy partners' portfolio and to become politically active in the energy producing regions.

1.1.1   Sketching alternatives: Southern Gas Corridor

Formation of the integrated global oil market in the late 1960s decreased vulnerability of the consumer states and diminished the asymmetric dependency. This had positive impact on the relationship between consumer and supplier countries. Suppliers become dependent on export earnings and upholding reputations for reliability on the background of shrinking market shares. Even in a tight international oil market or in the natural gas sector, where energy is not traded widely as a fungible commodity, there are strong counterbalancing pressures to import vulnerabilities that render interests interdependent among suppliers and customers (Stulberg, 2012). Consequently, these changes in the global energy market lead to the decrease of the cases with supply disruptions, which in earlier periods had serious damages on economy of partner states (Shaffer, 2009).

In the changing global energy market natural gas has started to be an important as oil. Especially in Europe, the shares of the imported natural gas grow rapidly, in the light of depletion of the domestic production and raise of the energy demand. The EU is interested in expanding the sources or the natural gas import and realization of the new pipeline projects from the Caspian region and the Middle East. However, the realization of natural gas supply projects is not an easy job. Political issues, conflicts, instability in transit countries and energy interests of Russia, which uses natural gas as political tool and wants to control natural gas supply from the Former Soviet Union (FSU) countries in the western direction, challenge the process.

Europe imports about 40 percent of the natural gas consumed by the European consumers from Russia through the Gazprom controlled pipeline and transmission systems, which traverse the territories of various transit states. The concerns over supply dependency have intensified following the natural gas supply disruption in January 2006, when a severe dispute arose between Russia and Ukraine over the gas prices, supply and payment. From 2006 to 2009 the gas crisis led to serious gas shortages in Central and Eastern European countries, especially during wintertime and demonstrated Europe's high dependency on a single supplier. The decline of the Europe's own hydrocarbon reserves in the Northern Sea and the escalation of the Ukraine-Russia relations observed with supply cut-offs have propelled energy security issues up to the top of the European Union's (EU) energy and foreign policy agenda. Diversification of the supply routes and energy sources, as well as development of the new transportation structures connecting Europe directly with other potential natural gas producing regions got high priority. Consequently the idea of the southern corridor (the Fourth corridor) started to receive much support from the European Commission (EC).

Initially, it was planned that the southern corridor would connect Europe with the Caspian region and the Middle East – primarily Azerbaijan, Turkmenistan, Iraq and Iran. In fact, existing conflicts and instability in the Middle East and North Africa have limited the ways of finding reliable sources of alternative supplies from these regions. Therefore, supply of natural gas from the Caspian has been considered as a milestone to start the realization of the Southern Gas Corridor (SGC), which would decrease EU's dependence on Russian imported gas.

Hydrocarbon resources of the Caspian basin have for a long time attracted attention of the foreign energy companies. The collapse of the Soviet Union created an opportunity for international oil and gas companies to start exploration works in the region. However, the landlocked nature of the Caspian region has composed a challenge by limiting energy transportation options. Transportation of energy resources from the region to the energy markets requires cross-border pipelines that transit territories of neighboring countries, and thus is a subject of increased risks. Hence, export pipelines are believed to be a key constraint both in terms of politics and the cost of exporting (Stevens, 2006; Stevens, 2010). Oil and gas transportation costs from the Caspian basin using pipelines are up to six times higher than in the other oil-producing regions of the world (Stevens, 2010). That is why the construction of cross-border pipelines requires large upfront investment, and the decision to build cross-border pipelines from the landlocked regions is influenced by political and economic considerations of all parties involved.

For the transportation of natural gas from the Caspian region several transportation routes and pipeline projects with different transmission capacities have been proposed including the NABUCCO pipeline, the Interconnector

Turkey–Greece–Italy Pipeline (ITGI), the Trans-Adriatic Pipeline (TAP), the White Stream and Azerbaijan-Georgia-Romania Interconnector (AGRI). All these pipeline projects were designed to bring SDII gas to Europe, which will provide the initial flow for the southern gas corridor.

It is worth to underline that among all these projects the NABUCCO pipeline was 'the flagship project of the diversification efforts of the EU in terms of security of supply. The EU was giving higher priority to this project in the opening of the southern corridor (Baev & Øverland, 2010).

However, various factors challenged the realization of the project. On the one hand, competing political interests of the regional and non-regional actors, on the other hand, commercial preferences of the producers and sellers, including energy companies, as well as energy policies of the consumer countries have added additional elements to the pipeline dynamics within the corridor causing frequent shifts and changes.

The current study focuses primarily on the pipeline politics and dynamics within the Southern Gas Corridor and provides an analysis of how energy policies of the state actors are shaped and under which conditions decisions are reformulated. The research explores the second stage of the Caspian energy development associated with the rise of the natural gas production and pipeline politics in the southern gas corridor.

The realization of the natural gas export in westward direction has become the part of the multidimensional pipeline politics. The factors affecting natural gas supply from the Caspian region differ from the oil export, which influenced the first stage of the Caspian energy development. In contrast to that first stage, the second stage of the is more complex, because of intertwined political and commercial factors, competing interests of the state and non-state actors. The decision-making was challenged by different issues. First, there are too many diverse players involved in the decision-making process. Second, following the successful implementation of the export projects during the first phase, producer countries became financially more independent and were not in search for FDI. They were playing a more active role in the decision-making than before and set their position mainly according to their foreign policy priorities. Third, energy policy objectives of the regional and non-regional powers have an impact on the decision-making process and the setting of the regional policy agenda. The energy politics pursued by major powers are capable to influence decisions of the smaller players and changing their positions. There are steadily emerging new conditions during the second stage, which are shifting the whole picture and the current pipeline dynamics. Finally, negotiation process becomes more complicated as interests and concerns of states and companies do not match.

1.2 Research design and methodology

1.2.1 Research objective, questions and hypothesis

The analysis of the pipeline and energy politics pursued at the second stage of the Caspian energy development requires complex analysis of interlinked issues. It should include both political factor (geopolitical constellation of the region, power relations and the level of interdependency among the actors, policy priorities of the key regional actors) and economic factor (energy market dynamics, commercial interests of the energy firms, nature of the natural gas supply and economy of pipelines) in order to get a clear picture of the energy and pipeline politics taking place in the region.

This study primarily focuses on the changing dynamics of the pipeline politics within the southern gas corridor and the factors affecting the decision-making process on the selection of the natural gas supply route from the Caspian region to Europe. In parallel, energy relationship between Russia and the EU, and Moscow's strategic interests in the region are to be reviewed in the research, since both have a direct impact on shaping the current pipeline politics.

The research period covers the second stage of the energy development in the Caspian region, starting from 2006 and focusing on the production and export of the natural gas to the European markets. The focal point of the policy analysis lies on the competition among the different pipeline projects targeting transportation of the natural gas from Shah Deniz full field production (SD II) and final decision of Shah Deniz consortium partners concerning the export route selection. The central question of the study is:

> What are the main factors affecting the current pipeline dynamics during the second stage of the Caspian energy development, and leading to delays in the realization of the Southern Gas Corridor, which is to connect natural gas resources of the region with the European markets?

This central question can be answered by addressing following sub-questions:

What are the key drivers influencing energy politics in the southern gas corridor?

- What are the key actors involved in shaping current pipeline politics and how do their political and economic interests influence the natural gas export route selection process from the Caspian region to Europe?
- How have the energy policy priorities of the littoral states changed throughout the second stage of the energy development and with what impact?
- How do market dynamics and preferences of the energy firms affect states' energy policy priorities? What are the advantages of the getting control over the corridor or supply chain?
- Which factors may lead to the failure or success of the strategic pipeline project? Under which conditions do the transformation of the pipeline project happen?

Since the collapse of the Soviet Union, oil and gas projects and perceptions of oil and gas potential of the Caspian Sea region have all been very strongly influenced by global and regional politics (Crandall, 2006; Stevens, 2006). The energy politics of the southern gas corridor involves many different stakeholders including seven states and eleven energy companies, whose interests coincide and differ at the same time. Moreover, the new gas corridor is designed in a way to change the energy map of the entire region and challenge the market power of the traditional actors, namely Russia. It is possible to identify two types of the actors engaged in the second stage of the Caspian energy development. On one hand, there are actors, who actively participate and are interested in realization of the supply corridor, but have certain preferences regarding the supply route. On the other hand, there are actors, namely Russia and Iran, whose political and commercial interests, as well as energy policy objectives will be challenged by the realization of the southern gas corridor. Therefore they are actively trying to prevent implementation of the huge pipeline projects from the Caspian region in westward direction.

Moreover, the states involved and expressed their interests to become a part of the pipeline projects of the southern gas corridor are classified according to their position within the supply chain: upstream, mid-stream and downstream. Three littoral states – Azerbaijan, Kazakhstan and Turkmenistan – are key suppliers constituting together a coherent regional energy system. There are regional transit states, namely Turkey and Georgia, which enable transportation of the resources from landlocked Caspian to the global markets and also compose significant elements of the energy corridor. With other words, supply network constructed during the first stage of the Caspian energy development has established a regional energy framework linking these actors (the supplier and the transit states) altogether through the new transportation systems. In fact, the cooperation among these states is based on compatibility of interests, where each however pursues different political and commercial objectives accordingly.

When analyzing the energy security objectives of the downstream countries, the fragmentation of the positions becomes more visible. Following the Ukraine-Russia gas crisis in 2006 and 2009, the EU by initiating the concept of the fourth corridor and providing political support for the NABUCCO pipeline project intended to reduce the economic and political vulnerability arising from import dependency, particularly in Central and Southeastern European countries. Even, if it is not evidently visible, the Russian factor play very important, as well as controversial role in the European energy and diversification policy. Despite the EU's diversification policy, some member states joined Russian backed pipeline projects[2] and were advocating some exemption rights for the Russian new pipeline projects, which lead to striking uncertainty and division amongst the European governments on energy related questions (Young, 2009). European energy security is highly dependent on diplomatic relations with Russia. Existence of several joint ventures between European and Russian energy companies has negative impact on European energy security, causing failure in development of the common energy and diversification strategy (Marácz, 2011).

The current pipeline politics illustrate a tough "struggle" among various pipeline projects and two different forms of competition between the major actors can be distinguished. The first one composes a competition between various pipeline projects within the southern gas corridor. Here ITGI, TAP, NABUCCO, SEEP, AGRI and White Stream struggle over the right to export Shah Deniz gas to Europe. The second one presents the rivalry between the regional powers and their initiatives, namely competition between Russia and European states, or Russian South Stream and EU backed Southern Gas Corri-

---

2   Bulgaria, Romania, Hungry, Serbia, Macedonia, Croatia, Slovenia, Austria, Italy, Greece and France became partners in the South Stream project and are willing to import more gas from Russia. The involvement of Nabucco countries thus are interested to reduce their import dependence from Russia the new pipeline project in the Russian South Stream project questions reliability of the partners and effectiveness of the European energy policy.

dor. All these increase uncertainty around the energy politics by turning the current pipeline competition into complicated game where actors play simultaneously together and at the same time, against each other.

Hypothesis I:

> Under the balance of power, the success or failure of a transportation project and shifts within the pipeline dynamics are determined by the interweaving moves of the all actors involved, rather than by the personal plans of a single one. The political and commercial interests of the actors shape the directions and the nature of the actors' moves.

The political and commercial interests of the actors represent an independent variable and shifts within the pipeline dynamics, including the final decision (success or failure of the project in terms of its realization) are dependent variable. The moves of the actors constitute intervening variable. The level of dependency among actors compose conditional variable. By the decrease of the power differences, actors lose the ability to control moves or decisions of each other. Since political and commercial interests are interlinked and the power relation among the actors may influence the decision making process, testing this hypothesis will show to what extent interests and changes within the power relations can lead to the shifts within pipeline politics.

The second stage of the Caspian energy development is reviewed with active participation of the littoral states in the energy policy making. During the first stage of the Caspian energy development newly independent states of the region had been highly dependent on foreign direct investments in order to be able to exploit their energy resources. The landlocked nature of the region was causing additional challenges in attracting investments. Hence, Caspian states were highly dependent on investors' decisions and policy preferences. The situation has changed when the oil pipeline projects started to bring revenues to the state budgets.

Formation of the strong political elite and revenue growth in Azerbaijan, Kazakhstan and later also in Turkmenistan created a condition, where these states became more actively involved in the decision-making process and put forward their own pre-conditions regarding the realization of new energy projects. In the early years after the collapse of the Soviet Union, external actors tried to maintain good relationship with traditional regional actors, namely with Russia and Iran, and their policy strategies were built on balancing the relations with both countries. Now, they have become engaged directly with newly independent states of the region.

Compared with the two Central Asian countries, Azerbaijan stays at the center of the debates around the southern gas corridor, which motivated central focus of the research. The discovery of the Shah Deniz gas field in 1999 shifted the perceptions of the natural gas supply from the Caspian region, and added a new dimension to the pipeline politics, and Azerbaijan became the key link in the natural gas supply from Caspian Basin towards European markets. That is, current energy politics in the Caspian region is characterized primarily through Baku's active participation in the decision-making processes and the realization of the pipeline projects.

As it will be seen, in the second stage of the Caspian energy development, regional states and energy companies are more successful in taking a stand for their commercial interests and priorities, than the EU for its political and strategic objectives. Based on this assumption, the following hypothesis is formulated.

Hypothesis II:

> Financial independence and the ability to attract funds for the realization of transportation projects, in case of passive participation of the bigger/consumer states, enable smaller states/supplier countries in the Caspian region to act more selectively in choosing supply routes and encourage these states to play a decisive role in the negotiation process, in order to maximize state revenues.

Here, financial independence and ability to attract funds constitute independent variable, whereas the way in which the supplier country behaves is the dependent variable. Testing this hypothesis will allow to conclude whether commercial preferences of the supplier countries may affect the realization of the strategic and politically important transportation projects. It is also worth to explore how relatively smaller regional player may affect energy policy directions of the other players and shape pipeline dynamics along the whole supply chain.

Furthermore it will be observed that, today Azerbaijan is positioning itself to be not only an oil and gas producer, but also a transit country for Central Asian energy resources destined for the European markets. Parallel to maximizing energy revenues, official Baku aims to strengthen its position as a major hydrocarbon exporter, to gain strategic leverage towards other players in the South Caucasus and at the same time to contain Russian and Iranian influences in the region (Rzayeva, 2012).

In actual fact, natural gas supply via pipelines from landlocked regions is a very complex process. At the same time, realization of the transportation from the Caspian region is easily challenged since it involves more than one pipeline project, which increases the numbers of the stakeholders. Furthermore, natural gas supply is more often politicized and has been used as a political tool or leverage by state actors (Correlje & Van der Linde, 2006), which affects commercial viability of energy production and realization of infrastructure projects (Shaffer, 2009).

Pipeline politics pursued within the southern gas corridor take place in a geopolitically complicated region. As energy politics around southern gas corridor take place in the light of EU's diversification policy and Russia-West rivalry, it adds another dimension to the pipeline politics by increasing geopolitical significance of the projects. In fact, actors involved in the energy competition pursue political and strategic interests towards the region as a whole, in addition to commercial and energy security objectives (Rzayeva, 2012). For the major actors in political and strategic context, control of the transportation routes ensures control of the resources (Amineh, 2003; O'Hara, 2004) and the market dynamics. It explains why, they are either interested in implementation or failure of the bigger strategic transportation projects.

The main game changer in the pipeline struggle was the announcement of the construction of Turkish-Azerbaijani Trans-Anatolia Pipeline project (TANAP), which replaced the entire part of Nabucco passing through Turkey, granting Azerbaijan and to some SD shareholders the access to the supply chain. TANAP was driven primarily by commercial considerations of the Azerbaijani government and energy companies involved in the Shah Deniz consortium. As a result, commercial interest of some actors prevailed over the strategic interests of other actors. In fact, along with political support commerciality and financial aspects of the pipeline project have a direct impact on the decision-making process. In order to analyze this aspect, the third hypothesis is postulated:

Hypothesis III:

There is a positive correlation between the economy of a pipeline and market dynamics, intention to increase the revenues and get a higher market share. As long as the pipeline project is bringing high rents to the partners along the supply chain, it has more chances to be supported by upstream states and business parties rather than politically motivated projects.

Commerciality of the project and the economy of the pipeline constitute the independent variable, whereas probability of the project to be realized is the dependent variable. Testing hypothesis will help to conclude in which cases market dynamics affects the decisions of the parties and commercial incentives prevail over political interests.

After analyzing different aspects of pipeline's commerciality it will be possible to find out under which conditions political and economic developments may lead to the transformation, failure or success of a pipeline project. Pipeline politics, in general, is a very dynamic game. Since it involves many stockholders, entering and exiting the game, changes are inevitable.

1.2.2   Current state of research

In recent years there has been a growing body of literature dealing with the energy security in general (Moran & Russell, 2009; Shaffer, 2009; Pascual & Elkind, 2010; Bahgat, 2011), the EU energy security (van der Linde, 2005; Stern, 2006; Young, 2009; Marin-Quemada, Garcia-Verdugo, & Escribano, 2012), energy politics in the Caspian region (Hopkrik, 1994; Ebel, 1997; Manning, 2000; Amineh, 2003; Dekmejian & Simonian, 2003; O'Hara, 2004; Gelb, 2005; Pomfret, 2005 & 2006; Mairet, 2006; Crandall, 2006; Bilgin, 2007; Fettweis, 2009) and pipeline politics in particular (Jentleson, 1986; McLellan, 1992; ESMAP, 2003; Stevens, 2000&2010).

The pipeline politics in Eurasia caused an intensive scholarly debate over the structural dimension of cross-border energy transit. Scholars state that export pipelines constitute physical-commercial ventures for moving oil and gas, which are subject to economies of scale, long lifecycles, large upfront investment, inflexibility, natural monopolies and the "tyranny of distance" in the case of natural gas (Stulberg, 2012). As a fixed infrastructure prone to market failure, the commercial value of a pipeline is directly affected by the dedicated upstream supply, price of the throughput, availability of alternative supply options and state intervention (ESMAP, 2003). Since transit pipeline passes through territories of various states, more than three stakeholders are involved in the process of pipeline construction, operation and rent sharing. These multiple stakeholders are left to their own devices to resolve conflicts of interests, protect vulnerable infrastructure, reconcile different national legal regimes and norms, and locate mutually rewarding outcomes for the reliable delivery of strategically important throughput (Shaffer, 2009).

According to Stulberg, cross-border pipelines are prone to crisis if parties fail to negotiate and sustain the terms for construction and operation. Therefore, pipeline politics refers to the unilateral and arbitrary disruption or renegotiation of the terms of supply, transit, off-take and delivery (Stulberg, 2012). However, it is possible to identify different levels of interdependencies between the parties along the supply chain, which also will influence the nature of the pipeline politics.

In general, oil and gas pipelines have been regarded as 'steel umbilical cords' of dependence, which can be disrupted for commercial and strategic gains (Ebel, 1997). Based on this possibility, scholars tackling with the pipeline politics developed various arguments regarding the nature of energy politics. Neo-realist and neo-liberalist approaches are applied to analyze pipeline politics in the literature. Neo-realists argue that the pipelines serve as instruments of competitive resource nationalism. Proponents of neo-liberalism and Marxism consider pipelines as conduits for constraining opportunism, strengthening interdependence and transforming interests in regional cooperation. Pipeline politics has been considered as a part of the national energy security, where 'the lands between' Russian/ Caspian suppliers and markets in Europe and Asia mere pawns in the global quest for energy security (Stulberg, 2012).

Neo-realists make an accent on the conflict-prone nature of the transit pipelines and consider them as a tool of power politics among the actors. Throughout history the significance of the pipeline has changed in terms of power and politics. When pipelines were transporting only a small fraction of the oil, it was traditionally considered as an issue of a tertiary economic concern and with little relevance to strategic behavior (Mearsheimer, 2001; Stulberg, 2012). With the rise of the amount of the traded oil and gas via pipelines, the extra-market value of the infrastructure also increased. A pipeline not only connects the supplier with the market, it also leads to the dependency between states along the connection line and can be exploited for strategic purpose.

Indeed pipelines are often part of resource nationalism politics of the states. The fact that more than 75 per cent of global oil and natural gas reserves, and that cross-border energy transit, falls increasingly under national authority speak for this impulse to control critical resource endowments (Gilpin, 1983; Hirschman, 1980). States with preponderant power are presumed to be especially inclined to struggle over pipelines control in their strategic orbit, given the sunk costs and asset-specificity of transit infrastructure (Frieden, 1994).

Control of pipeline routs gives power of influence to its owner in strategic regions. That is why regional suppliers and foreign investors are prone to engage in costly competition to control access to disputed fields and critical transit chokepoints in an effort to obtain influence over important regions and reduce the vulnerability of supply lines (Stulberg, 2012). On the other hand, the growing energy demand and market share of the energy resources force to see the pipeline politics from another angle. Hence, parties involved in pipeline politics also compete for the market shares. Scholars claiming that market competition will devolve inevitably to 'resource wars', relegating pipelines to service noncommercial foreign policy aims (Klare, 2002; Duffield, 2008).

Representatives of neo-liberalism see energy transit as an opportunity for a trade and cooperation. As pipelines create energy interdependence and provide mutual gains, neoliberals claim that it can strengthen cooperation between parties along the infrastructure. Disruption of the energy transit as in the case of the trade disruption will have negative effects on state at the macroeconomic level. The disruption will require time and additional costs for recovery. Hence, economic interdependence may constrain states in a positive way (Mansfield & Pollins, 2003).

Regarding energy security, supply stability determined by the ability to deter, mitigate and contain potential threats to the consistency of the delivery considered as a key factor. Market mechanisms, supply diversification, technological innovation and the availability of strategic reserves constitute 'shock absorbers' to attenuate the negative effects of price volatility (Stulberg, 2012). In this case, pipelines not only render the costs of conflict prohibitive in terms of disrupted supplies, but also provide instruments to soothe otherwise conflict-prone relationship among different parties (Fettweis, 2009). On the other hand, mutual economic benefits driven from energy interdependence and economic interests will deter partner states from deleterious economic effects of a breakdown of the energy trade (Gelpi & Grieco, 2008). In this case, democratic political institutions are seen as critical intervening variables, where increased economic dependence will reduce the trade disruption (Russett & Oneal, 2001).

There are studies proving that conditional and transformative strategies reinforce cooperation between neighboring and partner states (Stern, 2005; Kahler & Kastner, 2006). Globalization of trade and financial markets saps the capacities of states to advance pipeline projects on their own and it increases the availability of foreign direct investment, affording partners more opportunity to communicate true levels of resolve to realize mutual interests in energy transit (Stulberg, 2012). Since states are dependent on the private sector, their interests match on maximizing netback values and returns on investment (Jentleson, 1986).

While reviewing the current literature tackling energy and pipeline politics around the Southern Gas Corridor, it is possible to identify three main directions, which are somehow interconnected with each other. These frameworks include: a) geopolitical competition over the region and resources; b) EU energy security and; c) competition for market share.

1.2.2.1 First framework: Geopolitics

Many analysts referred to the concept of geopolitics to explain energy and pipeline politics from the perspective of zero-sum game played by the international actors in their pursuit of power and security. The geography and the landlocked nature of the Caspian region played a pivotal role in determining the direction of the politics of regional and non-regional actors. Since the realization of the expensive pipeline projects from the Caspian region has been the subject of the energy politics, it has been viewed from perspective of "division of rents" and geopolitical influence. Indeed, geopolitics of the Caspian region goes beyond just energy politics and has various dimensions. For some political actors, the region represents a scene where they exercise their powers and implement grand strategies towards each other (Brzezinski Z. , 1998; Buzan & Waever, 2003), where one party's gains necessarily create a loss for the other. As competition for power and influence were used to describe motives of the major actors involved in the region, a concept of "great game" in new formulation has been applied to explain energy politics pursed by these actors.

Following global power restructuring after the collapse of the Soviet Union, the Caspian region has turned to be at the center of international political debates. Hydrocarbon resources of the newly independent states of the Central Asia and Caucasus have been considered as alternative sources, which potentially will give to U.S. and European countries geopolitical advantages against Russia, Iran and China (Jaffe & Manning, 1999). During the 1990s, the Caspian region was sought to be a resource bonanza, a new "Persian Gulf", with competition over control of oil and gas reshaping geopolitics into a twenty-first century version of the "Great Game" (Manning, 2000).

Before, the term Great Game was used to describe "shadowy struggle for political ascendancy" (Hopkrik, 1994) that took place between the Russian and the British Empire in nineteenth century. In its nature, the old great game was a struggle for imperial dominance, power, security and control of territory. With the dissolution of the Soviet Union, political, military and economic situations in the region turned to be a fundamental part of the analysis and the main question of the "New Great Game". Consequently, the concept of the New Great Game has been used as shorthand for competition in influence, power, hegemony and profits, referring to hydrocarbon resources in Central Asia and Caucasus (Edwards, 2003).

The emergence of the new states in Central Asia and Caucasus, and the opening of the region for political and commercial actors, recalled the term again. This time the term has been used to describe not only struggle for influence and control, but at the same time, competition for economic profits from oil and gas revenues. Political dynamics have included geopolitics and geo-economics. Control of Caspian energy has turned to be a focal point of power in international politics. As it argued, the main prizes of the new game are energy contracts, pipelines, petroleum consortiums, and transportation routes (Meyer & Brysac, 2001).

Since the traditional aspect of the concept is political hegemony, policies determining geopolitical dynamics in the region have been interpreted either favoring or opposing Russian hegemony and influence in its 'near abroad'[3] (Weisbrode, 2001). Hence, competition over the influence in Central Asia and Caucasus between Russia and the Western allies became the integral part of the new great game. However, the South Caucasus due to its geostrategic location, namely being a bridge between Russia and Iran on one hand, and providing access to Central Asia through the see, was and stays at the heart of this struggle. As Cohen underlined, the real struggle for influence was taking place in Caucasus, where the presence of Russian military and the role it has played in the conflicts between and within Caucasian states. It was clear that Russia would not abandon its ambitions in the South Caucasus, on the background of increasing Turkish and American presence (Cohen, 2001).

---

3   Former Soviet territories were called as 'near abroad' by post-Soviet Russia, in order to show the link with former empire.

Despite political-military dynamics of regional politics, the greater focus has been on energy politics, which links political influence and economic influence. Since oil and gas remain critical for the global economy, the region's hydrocarbon resources and potential profit they can bring were at the center of political debates. The question of pipeline access to reserves – what route should they take, who should be responsible for their construction and safety, who charges and profits from them, composition of consortia and firms responsible for this – was composing a whole subsection of the New Great Game literature (Edwards, 2003). This new game included not only states, but also international companies, struggling among each other. Hence, among other factors, economic security and primacy linked directly with regions energy resources was the key element of the New Great Game.

In the study of energy and foreign policy priorities of the major powers, Mackinder's geopolitical concept, the "heartland theory", is broadly applied in the works of Sara O'Hara and Michael Heffernan. Two articles entitled "From Geo-Strategy to Geo-Economics: The 'heartland' and British imperialism before and after Mackinder" (O'Hara & Heffernan, 2006) and "Great game or grubby game? The struggle for control of the Caspian" (O'Hara, 2004) present energy politics of the major powers within the context of geopolitical rivalry. The authors argue that after the collapse of the Soviet Union the Caspian basin, due to its rich raw materials, represents considerable *prize* in itself. Gaining control over the routes by which oil and gas will be exported has been a crucial part of the struggle for control of the Caspian (O'Hara, 2004). According to O'Hara, Mackinder's heartland thesis stays at the center of the states' intention to control the Caspian, as the region has been included into the *pivot* area. By citing Mackinder, she added: *Who controls the export routes, controls the oil and gas; Who controls the oil and gas, controls the heartland* (ibid). For a major power control and the development of the resources, as well as transportation routes, have been regarded as a powerful combination of geostrategic and geo-economic calculations (O'Hara & Heffernan, 2006). USA and Turkey on one side, Russia and Iran on the other one, EU and China (in margin area) are identified as the main actors of this competition, where the last two have little influence, despite growing interests in the region.

A critical geopolitical approach to the energy security and pipeline politics in the region can be found in the works of Mehdi Parvizi Amineh. According to him, the interest of the regional and non-regional actors in getting control over the region can be explained by the growing energy demand, where the oil and gas reserves of the Caspian Basin constitute special importance for the energy security of the European Union (Amineh, 2003). Strategic geopolitical importance of the Caspian region reviewed along with Central Eurasian region[4] and the main focus has been done on region's geographical position, as being located between Russia, the Middle East and Asia-Pacific the regions play the role of a bridge. Considering involvement of the many actors in the region, Amineh classifies them through several levels (Amineh, 2004):

> The "inner circle" includes Russia, Iran, and Turkey, the "outer circle" includes (a) the more distant states China, India, Pakistan and also Afghanistan; and (b) the peripheral states Ukraine, Romania, Bulgaria, Greece, Ukraine, Israel, and Saudi Arabia. The United States, European Union, Japan and East Asian states are considered as external actors of the broader region. In fact interests of the transnational energy companies play also very important role in the formation of the energy politics.

For Amineh, the vast oil and natural gas resources of Central Eurasia have transformed the region into a location in which the forces of interstate rivalry, enterprise competition, and responses by regional state and non-state actors intersect (Amineh, 2003). Consequently, existing *multidimensional rivalry* increases uncertainty and unpredictability of the energy politics in the Caspian. In a situation of general political uncertainty, the security of gas supply specially becomes more insecure and economic uncertainty may trigger a supply shortfall in the future (Van der Linde, Amineh, Correlje, & Jong, 2004).

Application of the concept of geopolitics to explain and understand current energy politics, namely natural gas supply from the Caspian, is necessary, since geography and political choices of the actors direct gas trade along one route at the expense of another, investment and revenues are diverted as well, with considerable political implications (Victor, Jaffe, & Hayes, 2006). As soon as states become involved in a gas trade and commit themselves to import and export large volumes of the gas, their security concerns become interlinked. Hence, all partners are interested in maintaining the political stability of one

---

4   The collapse of the Soviet Union and the end of the Cold War led to a dramatic change in the landscape of Eurasian geopolitics. On the one hand, it resulted in the emergence of the eight independent states of **Central Eurasia**: the sub-region of Central Asia consisting of Kazakhstan, Kyrgyzstan, Tajikistan, Turkmenistan and Uzbekistan; and the sub-region of the South Caucasus consisting of Armenia, Azerbaijan and Georgia. On the other hand, it changed the control of the **Caspian Sea** basin from two littoral states—the Soviet Union and Iran—to the five countries of Russia, Iran, Azerbaijan, Kazakhstan and Turkmenistan. See: Amineh, M. P. (2003). Caspian Energy: A viable alternative to the Persian Gulf? *European Institute for Asian Studies , 03* (03), 1-20.

another. Consequently, the geopolitics of gas turns out to be a main issue discussed in this context. It's meaning is not limited to an endless struggle for global position, but also include the immensely political actions of governments, investors, and other key actors who decide which gas trade projects will be built, how the gains will be allocated, and how the risks of dependence on international gas trading will be managed (Victor, Jaffe, & Hayes, 2006).

1.2.2.2   Second framework: EU energy security

Reviewing the literature dealing with EU energy security issues and Caspian energy resources different, positions and approaches can be found. The EU interests and involvement in the Caspian region have been elaborated by two factors. First, instability in the regional states has been seen as a threat to Europe, due to its geographic proximity. Second, energy resources of the region have been considered as a valuable alternative, in the light of declining indigenous oil and natural gas reserves in Europe. The supply of the fossil fuels of the region, especially natural gas, to Europe started getting political meaning during the late 2000s, when Russian gas was taken as a key European security issue. We can identify three periods – before Ukraine-Russian gas disputes, intensification of the supply security debates between 2006-2013 and period after the selection of TAP as a priority route for the transportation of the natural gas from Azerbaijan – within the literature focusing on the dynamics of natural gas supply from the Caspian Basin to the European countries.

At the beginning of the 2000s, energy supply from the Caspian Basin was evaluated mostly from an economic perspective. In this context, a series of studies published by the research team at Clingendael International Energy Programme (CIEP) is noteworthy, since the studies provide deep analysis and broad perspective on the changes of the EU energy security within the geopolitical framework. According to that research, the EU energy security concerns emerge from the set of developments of geopolitical and economic origins (Van der Linde et. al., 2004; Correlje & Van der Linde, 2006). While analyzing security of natural gas supply, initially, supplies from Russia and future supplies from Caspian Basin through the Russian transmission networks was seen as a necessary step towards fulfilling growing energy demand in Europe. Moreover, reserves from the Caspian should not be considered as an alternative to the Russian gas, but as an additional source for the European market (Van der

Linde, Amineh, Correlje, & Jong, 2004; Correlje & Van der Linde, 2006). Similar to Correlje and Van der Linde, a scholar from the Oxford Institute for Energy Studies, Jonathan Stern, underlined non-feasibility of a quick shift away from Russian natural gas supplies (Stern, 2002).

However, this perception has changed when European countries started suffering from supply disruptions as a result of the Russian–Ukrainian gas disputes of 2006 and 2009. Besides, a wide range of analysis, academic and political, paid particular attention to the politicization of the gas trade, secured gas transit and transit risks. Diversification of the supply routes and sources became the priority issues within the EU energy security and political framework. The authors tackling the energy security issue from the political perspective consider dependency from Russia as threat to the EU energy security and question's its reliability as a stable natural gas supplier. Diversification of the natural gas supplies turned to be foreign policy and energy security priority in the European agenda following natural gas shortfalls experienced during the disputes between Ukraine and Russia, where increased Europe's dependence from Russian supplies become more tangible. On the other hand, some authors do not see dependence from Russian sources as an issue at all and mainly focuses on the transit risks driven from political instability mainly in the transit states. This group of scholars focuses on the issue from the commercial perspectives and elaborate different alternatives for the secured market functioning.

In the context of source diversification, importance of the development of new supply routes initially from Azerbaijan and later from Kazakhstan and Turkmenistan is underlined in the analysis of Mankoff. According to him, the strong state control of the natural gas production and its use as political leverage by Moscow will question reliability of Russia as a supplier (Mankoff, 2009). Considering the fact that for the time being the relationship between Moscow and Brussels within the framework of the gas trade is defined as interdependent, the situation will tend to change with the construction of the new pipeline systems from Russia and European countries. For Mankoff increased proportion of Russian gas consumed in the European countries along the new pipeline routes[5], possibility to cut supplies in current transit countries – Ukraine, Belarus, and Poland through bypassing their territories will increase, Moscow's political and economic leverages in these countries (Mankoff, 2009).

---

5   Here author considers two pipeline projects: Nord Stream and South Stream.

Indeed, it will have different consequences for the different EU member states and lead to the split of the positions regarding the supply security concerns within the EU. However, for official Brussels realization of the southern gas corridor and the Nabucco project was central to any discussion of diversifying Europe's energy supplies. As Mankoff states, the main problems ensuring supplies from the Caspian include weak political influence of the EU in the region, persistence of Russian interests and low engagement of the Central Asian suppliers in the project. On the other hand, while reviewing "In-depth study of European Energy Security", accompanying the document Communication from the Commission to the Council and the European Parliament: European energy security strategy, the main focus of the natural gas supply from the Caspian region is given to Azerbaijan rather than to Central Asian natural gas reserves.

By contrast, one may find totally different points regarding the significance of the southern gas corridor for the EU energy security in Pierre Noel's article. He argues that considering current changes of the European gas market and development of LNG transportations, the realization of the multi-billion euro merchant pipeline project from Central Asia, Caspian through Turkey to Europe is the wasted time and energy (Noël, 2013).

Similar to Noel, Pavel Baev also emphasize the changed approach to the energy security problems in Europe with development of LNG transportation and 'shale gas' revolution in USA. For Baev, economic underpinning of fundamental decisions made in the European Commission regarding the liberalization and diversification of the energy supplies has extremely changed in recent years (Baev, 2012). He also describes the EU's energy politics pursued as a part of securitization after mid-2000s in as *energy-geopolitical game*, which comes to the end.

Analyzing political aspects of the southern gas corridor, within the context of growing environmental concerns and failed gas diplomacy between the EU and Russia, in the research paper "From European to Eurasian energy security: Russia needs and energy Perestroika" the author argues that opening of the fourth corridor can be found in the interplay of political ambitions. He continues stating (Baev & Øverland, 2010):

> The idea of increasing the inflow of 'new gas' went clearly against the ideology of reducing the consumption of hydrocarbons but was nevertheless embraced by the European Commission as the materialization of the diversification guideline. The key asset in this corridor was supposed to be the Nabucco pipeline as 'an embodi-

ment of the existence of a common European energy policy. The project enjoys the most favored status in the EU Commission, but the consortium of six energy companies has been unable to get it off the ground[6].

Moreover, the main problem of the Nabucco's failure for Baev, should not be seen in the high cost of construction, but in the scarcity of supply sources, since the Shah Deniz gas would not fill a half of the planned pipe and European leaders were not successful in convincing Turkmenistan to join and give commitment to the project (Baev, 2012).

Pipeline competition between the EU and Russia intensified with Moscow's enthusiasm to open a new route to the European market through constructing the South Stream pipeline across the Black Sea and engagement of the European energy firms into the new project. According to Baev, both projects faced with the delays and transformation as a result of the financial crisis in Europe. As long as financial question and reliability of the sources stay as an issue, geopolitical' pipelines will re-categorized as far-fetched extravaganza (Baev, 2012).

Along with the studies of several scholars, there are also a huge number political papers focusing on EU energy security, diversification policy and natural gas supply issues. Starting from 2006 the significance of the Southern Gas Corridor and supply of the natural gas from Caspian region is underlined in the Communications of the EU Commission. The construction of the corridor was listed among the six priority infrastructure projects, which will supply EU's future needs. In the EC communication it is highlighted:

> The southern gas corridor is one of the EU's highest energy security priorities. The Commission and Member States need to work with the countries concerned, notably with partners such as Azerbaijan and Turkmenistan, Iraq and Mashreq countries, amongst others, with the joint objective of rapidly securing firm commitments for the supply of gas and the construction of the pipelines necessary for all stages of its development (Commission of the European Communities, 2008).

Besides engagement of Azerbaijan, involvement of Turkmenistan and other Central Asian countries as main supplier countries and the achievement of the agreement on Trans-Caspian Gas Transmission and Infrastructure between the

---

6   For more elaborate analysis, see Baev, P., & Øverland, I. (2010). The South Stream versus Nabucco pipeline race. *International Affairs*, *86*, 1075-1090.; and debates on a common European foreign policy on energy in the European Parliament at http://www.europarl.europa.eu/ sides/getDoc.do?pubRefl/4-//EP//TEXT+CRE+20070925+ITEM-015+DOC+XML+V0//EN

EU, Azerbaijan and Turkmenistan has been illustrated as an important element of the security of natural gas supply in the official documents of the European Commissions. In the Communication on 'Energy infrastructure priorities for 2020 and beyond', the Commission has outlined a master-plan for an integrated energy network taking into account key interconnections with the regional countries (Commission of the European Communities, 2010). Also, EU's active involvement in the process is stated in a Communication of 2011:

> The leverage of the EU internal energy market should be better used to facilitate large-scale infrastructure projects linking the EU network to third countries, particularly ones with political, commercial or legal uncertainties... Negotiating mandates for the EU may be necessary where agreements have a large bearing on the EU energy policy objectives and where there is a clear common EU added-value. The recent adoption by Council of a mandate to authorize the Commission to negotiate an agreement for the legal framework with Azerbaijan and Turkmenistan for a Trans-Caspian gas pipeline system offers an immediate example of the benefits of EU-level action for energy security (Commission of the European Communities, 2011).

Re-realization of the new transportation infrastructure has economic and political priorities for the EU. Brussels aimed to enable *a physical access to at least two different sources for every European region* (Commission of the European Communities, 2010), to ensure natural gas supply security. The political engagement of the EU directly with Azerbaijan and Turkmenistan and support for the Trans-Caspian Gas pipeline as a part of its diversification policy was negatively viewed from Moscow.

Valentina Feklyunina in her article entitled *"The 'Great Diversification Game': Russia's Vision of the European Union's Energy Projects in the Shared Neighborhood"* examines Europe's diversification policy from Russia's perspectives. She argues that although the EU is not yet seen as a serious threat to Russian interests in the area, this situation is rapidly changing, with the Kremlin becoming increasingly sensitive about the EU's plans to diversify energy supply sources and transportation routes by increasing cooperation with other former Soviet Republics within the Commonwealth of Independent States (Feklyunina, 2008). Since the EU actions in the Caspian region are viewed as an anti-Russian, Moscow mobilized its power and financial resources to maintain its position not only in the European markets, through diversifying its export routes, but also to prevent the realization of the huge transportation infrastructures from the region. According to Feklynunina, the problem becomes more complex as Brussels and Moscow have contrasting visions of what constitutes 'energy security', which also cause division of the views within the EU (Feklyunina, 2008).

While reviewing literature on the EU's supply security and diversification policy it is possible to define two developments. First, both the EU and Russia become involved in a 'diversification race' in the pipeline projects – South Stream and Nabucco – that are in a direct competition with each other. Second, European-Russian competition led to the politicization of the southern gas corridor initiative. In the light of these developments, it is worth to analyze interests and decision making of the regional countries, which will commit huge volumes of the natural gas for transportation.

1.2.2.3 Third framework: competition for market share

Reviewing the literature on energy security, it is possible to see that there is a tendency to focus on the geopolitical dependencies of smaller states and the power of larger states to determine economic and political outcomes (Goldman, 2008). However, sometime smaller regional states may play a more crucial role than major powers. Literature covers energy and foreign policy priorities of the upstream and midstream countries along the southern gas corridor. The broader attention in scholarly work has been given to the pipeline competition within the southern gas corridor, and the political and commercial interests of Azerbaijan and Turkey in the analysis of the Center for Strategic Studies in Azerbaijan and Eurasia program of the Jamestown Foundation. When the memorandum of understanding (MOU) between Azerbaijan and Turkey on the construction of Trans-Anatolian pipeline (TANAP) was signed, in December of 2011, a number of studies provided different explanations of the decision and implications of TANAP for the whole initiative. This decision has been considered as a game changer of the pipeline politics within the southern corridor, which also underlined the strong influence of the down-stream and mid-stream countries' interests.

The most detailed analysis of the TANAP project can be found in the works of Socor, Cain, Ibrahimov, Bilgin and Rzayeva. By replacing the Turkish part of Nabucco, the new project not only has sealed Nabucco's fate, but also provides a direct access to the market for the downstream stakeholders. Cain, Ibrahimov, and Bilgin in the article "Linking the Caspian to Europe: Repercussions of the Trans-Anatolian Pipeline" state that TANAP emerged as the prefer-

red pipeline to Europe from the Caspian, because of its local political and economic appeal, which also implies that regional politics when combined with
commercial interests and local market development can trump geopolitical
resource competition (Cain, Ibrahimov, & Bilgin, 2012). According to the authors, other reasons – predominantly economic considerations – rather than
geostrategic power politics were more crucial for achieving the intergovernmental agreement (IGA) between Azerbaijan and Turkey. Besides this
scholars argue that in the light of disagreements among the Nabucco partners,
TANAP turned to be an attractive investment vehicle on the one hand, and will
have significant economic and social implications for Azerbaijan, Georgia and
Turkey similar as Baku-Tbilisi-Ceyhan pipeline, on the other hand.

Gulmira Rzayeva, in her analysis explains the significance of the various
routes for Azerbaijan's energy security and policy priorities (Rzayeva, 2010 &
2012; Rzayeva & Tsakiris, 2012). In contrast to Cain, Ibrahimov, Bilgin, she
does not exclude political factors, which also affect the decision-making process. According to her, decision on the supply of natural gas to Europe is affected by commercial and political considerations of all the stakeholders. However, Azerbaijan stays at the center of these debates as a main energy producer
and transit state (Rzayeva, 2010; Paul & Rzayeva, 2011). Examining complex
issues around the production sharing agreements and decision making process,
she argues that the current situation within the southern gas corridor demonstrates that the issue of controlling the strategic infrastructure along the value chain
is becoming increasingly important for the stakeholders and is leading to
an implicit rivalry between the partners of the Shah Deniz II project in ac-
quiring the majority stake in midstream projects (Rzayeva, 2012). Further elaborating on the transformation of the initially suggested pipeline routes and the
development of the new initiative, namely the TANAP project, Rzayeva underlines the importance of balancing Azerbaijan's commercial interests with its
political priorities, foreign and domestic.

Turkey, as one of the region's key actors, gets further attention in the
scholarly analysis (Winrow, 2004; Bacik, 2006; Bilgin, 2007; Rzayeva, 2014),
due to being one of the important links of the energy corridor connecting the
sources with the market. Its geographic position makes this country significant
both from the political and economic point of view in the international energy
arena. Mert Bilgin and Gareth Winrow review the policy priorities of Turkey
regarding the southern corridor in a broader context. Economic, political and
transportation security of energy resources through Turkey is highly linked

with the EU energy security (Bilgin, 2007). According to Winrow, Europe's long-term energy security needs could be met if Turkey becomes a key gas transit state (Winrow, 2004). In fact, both scholars underline the importance of the realization of other pipeline projects for Turkey's strategic ambitions.

Through applying geopolitical concept scholars elaborate Turkey foreign policy in the region. In the article "Tangled Pipelines: Turkey's Role in Energy Export Plans", elaborating on Turkey's aspirations to be an energy hub for oil and natural gas exports from the Caspian Sea region and beyond, Carol Saivetz goes on to discuss the complex geopolitics of the region. She argues that Turkey must balance its aspirations to be central to regional export schemes and its increasing energy dependence on Russia (Saivetz, 2009).

According to Rzayeva, Turkey is prioritizing its own interests – to secure gas for its own market and to pursue its aim of becoming a hub. These interests are not always in line with EU policies on the realization of the fourth energy corridor project. Focusing on the Turkish natural market, the author explains that the willingness of Ankara to transit Caspian natural gas through its territory is also driven from natural gas demand within the country. In the long run, she states that Ankara plans to absorb most of the gas volumes available for export from Azerbaijan (Rzayeva, 2014). In fact, Turkey is representing a lucrative market already with a high netback margin for the Shah Deniz II partners, because of the short transportation distance and prices close to the European average price. As it can be seen, Turkey's active involvement in the negotiation process over the transportation route within the southern corridor with the regional supplier countries (Azerbaijan and Turkmenistan), is mainly determined by its willingness to become an energy hub. However, this motivation does not include only commercial considerations, but goes beyond and also involves political and strategic interests.

The analyses developed by various scholars and presented above are used for constructing an analytical framework and the explaining a more complex picture of pipeline politics taken place in the southern gas corridor. In general, fragmentation of the study frameworks and persistence of one-side approaches illustrate the existing challenge and the gap in the research of current pipeline politics. A more detailed explanation is given in the following part.

1.2.3   Conceptual challenges and research gap

Despite the existence of numerous literatures covering energy security, oil and gas politics in the Caspian region and pipeline politics within the southern gas corridor, there are still some challenges and gaps within the analytical framework. These are crucial for to answer research questions raised with the dissertation. In order to explain why and how intertwined political and commercial factors shape and reshape the current pipeline politics in the southern energy corridor, in-depth analysis is needed to determine the theoretical limit and the empirical gaps.

While reviewing the relevant scholarly and political analytical works the following shortcomings have been identified. First, by applying the concept of geopolitics, some authors do not differentiate sufficiently between oil and gas politics pursued during the first and the second stages of the Caspian energy development. Both stages have been introduced as part of the New Great Game between the great powers, driven from their power and political ambitions. However, the political landscape and the economic situation have tremendously changed in the Caspian region. Regional, relatively smaller powers start playing a more active role in decision-making process now, rather than before.

Second, economic interests and commercial considerations of the key stakeholders have been considered imperceptibly in the analysis of the selection of the transportation options. Natural gas supply from the region to Europe was mostly politicized, leaving little space for application of theories focusing on economic aspects of the actors' behavior, on pipeline economics in particular and geo-economics in general. Analysis of the economic indicators are mostly based on quantitative analysis illustrating energy indicators of reserves, production, demand and supply in numbers. Conversely, complex analyses of both economic and political factors are required to get a clear and comprehensive picture of the situation.

The pipeline politics (or pipeline race) within the southern gas corridor were viewed as a part of a strategic game between the states, whilst business interests of the energy firms and state-firm relations were not particularly elaborated. However, energy firms' decisions to enter or exit a competitive project influence the reliability of the proposed pipeline project per se.

Finally, the concept of interdependency was rarely used to explain the decision-making process. Evaluation of energy security leads the way to the analysis of interdependency and dependency between supplier and consumer countries emerged as a result of pipeline construction. Other forms of interdependency and theoretical frameworks were mostly not applied. In order to understand why regional suppliers, like Azerbaijan and Turkmenistan, need to counterbalance with the interests of Russia, why Moscow is not against of TANAP, and what changes have taken place after the financial crisis have to be correctly identified. To numerous explanation of the pipeline dynamics of the southern energy corridor requires a complex approach of political and economic interests of the state and non-state actors.

1.2.4   Structure of the dissertation

This thesis is structured in seven parts. Chapter 1 provides a general background for the natural gas security and pipeline politics pursued in the Caspian region from 2006 onwards. In addition, this part helps to understand the problem and gives an overview of the research and insight into the research framework need for this thesis.

Chapter 2 constitutes the theoretical framework of the thesis. Elias's concept of figuration, Strange's theory on structural change of economy, and Mercille's concept of radical geopolitics are utilized to explain the energy politics and pipeline dynamics from theoretical approach. The application of these three different concepts together contributes to the understanding of how dynamics shift and why policies are changing.

Chapter 3 elaborates the different dimensions of the energy security studies and the challenges related to pipeline transportation in general. The overall objective of this part is to present the link between the energy security and fossil fuel shipment from the landlocked and remote areas, and also analyze certain aspects of pipeline economics.

In Chapter 4 describes energy politics and interests of major regional actors during the first phase of the Caspian energy development. Stakeholder analysis and political impediments presented in this part help to distinguish differences between the various factors affecting the dynamics of pipeline politics pursued starting from mid 1990s and continued during the 2000s.

Chapter 5 elaborates the concept of the Southern Gas Corridor, its impact to the EU's energy security and provides a comparative analysis of the pipeline projects proposed within the new supply corridor. This chapter focuses on energy politics around the southern gas corridors and analyzes pipeline projects proposed for the transportation of the gas. Presenting the key figures and numbers, especially supply-demand relations, potential natural gas production in the Caspian region, the chapter analyzes to what extent Caspian natural gas production can meet Europe's energy demand and ensure security concerns.

Chapter 6 then presents a comprehensive stakeholder analysis. Interests and measures taken by the key actors in order to achieve their energy policy objectives in the region are described. A special focus is given on development of the TANAP project, on Azerbaijan's and Turkey's energy policy priorities, and on the interests of the energy firms stipulated with the implementation of the southern gas corridor. The last chapter also includes a summary of the thesis and presents the main conclusions.

2 Theory and Methodology

2.1 Theoretical framework

The aim of this chapter is to elaborate a theoretical framework and to present conceptual tools used to analyze the dynamics of energy security and pipeline politics in the Caspian region. For this purpose, three interdisciplinary and complementary concepts are applied to structure the framework: the concept of figuration; structural change of world economy; and radical geopolitics. Application of these different concepts together contributes to the understanding of pipeline dynamics and current energy policy pursued by various actors in the southern gas corridor. They address not only "how" dynamics shift, but will help to explain "why" policies are changing.

The figuration concept introduced by Norbert Elias was developed on the assumption of interdependence formed as a result of 'relation among actors and causality of actors' moves. By focusing on conditions and dynamics of various game models, the figuration concept explains why power balances change among players within the network of interdependencies, on one hand, and constrain or empower a particular player, on the other hand. The merits of the application of the concept of figuration to the pipeline politics are threefold. First, it focuses on relations and interdependences within a particular figuration. The interdependency among actors limits their actions and determines their moves within a particular figuration. Second, it presents various game models ranging from simple to more complex ones and uses different tools to analyze the dynamic nature of power and the ways of how it varies among the actors. The third feature of the figuration concept is the possibility of unintended outcomes happening within a particular figuration because of presence of other variables apart of actors' interests and plans. Application of the figuration concept may help to understand the dynamic and the logics of the moves of actors involved in pipeline politics.

However, without identification of the actors and the nature of the interdependencies it will be hard to provide exact explanations to the moves and inte-

© Springer Fachmedien Wiesbaden GmbH, part of Springer Nature 2018
S. Amirova-Mammadova, *Pipeline Politics and Natural Gas Supply from Azerbaijan to Europe*, Energiepolitik und Klimaschutz. Energy Policy and Climate Protection, https://doi.org/10.1007/978-3-658-21006-9_2

rests of actors. In order to understand the current political and economic dynamics taking place within energy politics it is worth to also look to the concept of structural change of world economy by Susan Strange, where she elucidates the mechanisms of change in political economy. The concept is not limited to the analysis of the process at the global level and but is equally useful for explaining dynamics of interactions between state and non-state actors.

The third concept used to conceptualize the energy politics is radical geopolitics. This concept developed by Julien Mercille offers a critical approach and incorporates political economy into the process of the analysis to examine the causes of state policies and political events on the international level. Radical geopolitics examines the importance of geopolitical and economic factors that drive policies of states. The advantage of the application of radical geopolitics while analyzing pipeline politics is that it not only focuses on geopolitical aspects of the policies, but also includes ideas from political economy. In fact, political factors together with economic forces play a crucial role in shaping states' policies. In order to be able to identify the reasons behind the foreign policy of international actors more clearly, geopolitical and geo-economic factors should be considered as mutually interdependent. This creates a condition, where foreign policy of major actors involved in regional politics, as well as energy politics can be interpreted. Moreover, applying two forms of logics to analyze certain events in the international arena helps to understand the interests of diverse groups of actors, and the rationale behind competing interests of major powers involved in energy politics in the Caspian region.

2.1.1 Figuration

The purpose of this part is to explain the structure of interdependencies formed as a result of pipeline networks from a sociological perspective by applying Norbert Elias's concept of figuration. In Elias' concept, individuals and society constitute the key units of analysis. These two units are strongly linked so that it is impossible to imagine both existing independently and separately from each other. If concepts of individuals and society will be considered as detached, then the society could hardly be conceived as anything rather than a collection of windowless monads (Elias, 2000). For Elias, the real condition of the relations is exact of converse among individuals (units), which at the same time forms social structure and social processes. Therefore, individuals cannot be considered as isolated, detached from others, since all are interdependent in different ways. Such interdependencies are the nexus of what he called figuration, a structure of mutually oriented and dependent actors. Moreover, within a particular figuration actions of actors are influenced and constrained reciprocally (Elias, 2000).

To explain the concept figuration as a system, Elias offers the metaphor of a dance, be it a tango, a mazurka, a minuet, a polonaise or rock 'n' roll.

> One can certainly speak of a dance in general, but no one will imagine a dance as a structure outside the individual or as a mere abstraction. The same dance figurations can certainly be danced by different people; but without a plurality of reciprocally oriented and dependent individuals, there is no dance. Like every other social figuration, a dance figuration is relatively independent of the specific individuals forming it here and now, but not of individuals as such. It would be absurd to say that dances are mental constructions abstracted from observations of individuals considered separately (Elias, 2000).

Figuration can thus be viewed as a moving picture where actors and their actions are united and constitute complementary parts of the web of interdependencies. Stating it with other words, the move of one will make sense, if it is analyzed together with the moves of other actors.

In fact, the figuration itself is not static. It is dynamic and tends to change. Hence, even small shifts in actor's move will lead to a change within the whole figuration. Such changes do not have causal explanations. A change in a figuration is explained partly by the endogenous dynamics of the figuration itself and the immanent tendency of a figuration of freely competing units to form monopolies (Elias, 2000). The concept of figuration can be applied to analyze not only dynamics of small figurations, but also larger and complex ones such as states, regional configurations and broader international systems.

A central point of the figuration concept is interpretation of power and power relations within the network of interdependencies. Here, the concept of power is presented as a relation rather than as a thing. Power is not used by specific people or groups against others, but is type of social configuration in which all actors are involved (Burkitt, 1993). Elias reviews power as a structural characteristic of all kind of relationships and underlines that "power is not an amulet possessed by one person, and not another" (Elias, 1978). Stating it differently, each human relation takes place in the framework of power relation, where none has an absolute power over others. Consequently, in a figuration no one is ever without some element of power, not only to constrain the actions of others but also to determine their own moves (Burkitt, 1993).

The presence of variables in terms of resources or a function may affect the social configuration by increasing advantages of certain groups of actors compared to others. These factors provide a greater opportunity for some actors to influence actions and determine their own moves. However, the relative power of different groups or classes still rests on the wider figurational network rather than on the ownership of resources or the performance of a function (Burkitt, 1993). The relevance of resources as well as the chance to increase influence or dominance depends on the function that an actor or a group has in a web of social interweaving (Isachenko, 2012). In addition, reviewing place of these factors in the context of changing pattern of power balances will help to better imagine the whole picture of relations.

Power like a figuration is not static and may change together with figurational change. Fluctuating nature and relativity of power is better explained through terms of mutual dependency and function. In all forms of figuration, ranging from simple to large, all actors are usually tied with each other in a tensile equilibrium caused by mutual dependency. Referring to relation between parents and child, as well as to Hegel's master and slave dialectic, Elias argues that despite uneven distribution of power and variation of power differentials from small to large, balance of power is always present wherever there is a functional interdependency (Elias, 1978). Stating it differently, one may conclude that because of power balances, even power chances are disseminated unequally in both models there remains a mutual dependency between the actors. Therefore, an actor with little power chances also retains a certain degree of power over another one.

The concept of function similar to the concept of power is conceptualized as a relationship in the figurational analysis. Here the concept of function refers to the interdependency of actors, where a function of one actor cannot be understood without taking into account other actor's function. Elias clarifies the concept of function and at the same time explains the causes of interdependency between actors of certain figurations in the following form:

> "...when one person (or group of persons) lacks something which another person or group has the power to withhold, the later has a function for the former. Thus men have a function for women and women for men, parents for children and children for parents. Enemies have a function for each other, because once they have become interdependent they have the power to withhold from each other such

elementary requirements as that of preserving their physical and social integrity, and ultimately of survival" (Elias, 1978).

The concept of function is linked with the concept of power and should be reviewed in a similar context of relationship among actors. The actions of actors in a particular figuration are linked. Because their actions are interdependent, like in the chess game, where move of one determines move of another one (Elias, 1978). Hence, in every kind of figuration the sequence of players' moves has to be interpreted through the process of interweaving and interdependency, which considers every move as consequent of the previous one.

The concept of power, due to functional interdependencies, becomes more accurate if it is analyzed in terms of power balances, power ratios and power differentials (Isachenko, 2012). Power as an integral element of relations and functional interdependencies does not exist apart from social configuration and like human relationships in general, it is either bi-polar, or multi-polar (Elias, 1978). Similarly, balance of power is either bi-polar or multi-polar, based on form of specific figuration, be it simple figuration with two actors or more complex one, with many actors. By presenting various models of games ranging from a simple to more complex, with several players and levels, Elias explains how the dynamic of power and the outcome of game may change because of these three factors (Elias, 1978). Moreover, the distribution of power among players or differences in power potential of each player play a significant role in shaping the dynamic of the relation within the particular figuration, as it illustrated in different game models.

In the framework of the game model with two players, the player A with relative higher power potential have better chances to influence decisions and actions of weaker player B. However, it does not mean that A has an absolute power over B. Since no player wielding absolute power, there cannot be player with zero-degree of power (Elias, 1978). Here the power difference between A and B is referred to power ratio and uneven balance, where that difference "determines to what extent player A's moves can shape player B's moves, or vice versa" (Elias, 1978). Hence, the ability to control the actions of relatively weak player, in this case, also provides an opportunity to a stronger player to control the dynamics of the game. Here the changing figuration will be more dependent on aims and plans for the course of the game formed by stronger player.

In contrast to this, in the game model, where balance of power is equally distributed among players, players have fewer chances to control actions of their opponents, the course of game and the changing figuration. Consequently, by decrease of power differences the dynamic of the game will change independently from personal plans of the players and will be result of the interweaving of moves (Elias, 1978). That is why power is a matter of balance and degree. As Elias articulates it, the moves of certain players can be limited and constrained by others, but it is a matter of degree, which results from their relatedness and their interdependency (Elias, 1978).

The process of figurantional change in complex game models, to some extent, differs from simple one. In a game model with multiple players, where players act separately and lack coordination of their actions against the single opponent, the trajectory of the game is similar to a game model with two players. Of course, there are also game models, where weaker players unite and play a single game against a stronger player. In this case, despite power differences, the dynamics of the game (or odds of one to control moves of others) are determined by the level of inner tension among the weaker group of players. The greater the tension, the greater is the chance of the former to control the moves of the latter, as well as the general course of the game (Elias, 1978).

More complex game models not only include many players, but also can be multidimensional and take place in two or more levels. The figuration of such games is very variable and mixed due to sequence of moves of players. Therefore, a player to be able to figure out its first and next steps in order to get the wished outcome, it needs to have a clear picture of the course of the game and of its general figuration, which changes constantly as the game proceeds (Elias, 1978). However, this is not an easy task even for a stronger player. By the rise and entrance of the new players into the game the course of the game and pattern of figuration become more and more uncertain, and at the same time almost impossible to control. Change within the figuration shifts power balances, on the one hand, and player's perception, on another one. Due to all multiple interweaving, it becomes more difficult for those actors to determine the next moves correctly and get desired outcomes.

Constantly changing figuration may lead to reorganization of players into different groups. This will be the next level of the complex game model. The next level may tremendously vary from the first one, due to entrance of new player, reorganization of old players in different groups, changes in time and space, power balances and power differences. Moreover, the power differentials among the players of specific levels will be not the same. Even within this form of figuration, the levels of game are interdependent and possess different reciprocal power chances corresponding to the degree of their dependence on each other (Elias, 1978). In order to explain changes and shifts within the figuration, one must focus on all changes and shifts within a figuration as inseparable dimensions and part of unique dynamic. Furthermore, the change cannot be explained as a set of causal actions and motives, as long as the change itself is the result of functional interdependence (Elias, 1978).

Game models with many players and several levels will have possible results of unplanned outcomes, even though the actions are intentional (Isachenko, 2012). According to Elias:

> From the interweaving of countless individual interests and intentions – whether tending in the same direction or in divergent and hostile directions – something comes into being that was planned and intended by none of these individuals, yet has emerged nevertheless from their intentions and actions. And really this is the whole secret of social figurations, their compelling dynamics, their structural regularities, their process character and their development; this is the secret …of relational dynamics (Elias, 2000).

Despite motives and actors' intentions play a crucial role and affect the dynamic of figuration, the dynamic of the process, cannot be explained only in terms of actors' motives and intents. There can be many factors and variables affecting the process. As Elias explains, "figuration had to arise out of certain earlier figuration, but does not assert that the earlier figurations had necessarily to change into the latter one" (Elias, 1978; Walsh, 2013). This leads to a crucial implication that there can be explanations, which are not necessarily based on causality, because changes in the pattern of figuration may be, in fact, occurring due to changes of the internal dynamics of a figuration itself (Isachenko, 2012). That is why, move of one player can only be interpreted in the light of the way the preceding moves of two players have intertwined, and of the specific figuration which has resulted from this intertwining of individuals (Elias, 1978).

The application of the concept of figuration, thus is not limited to focus either on individual or society, it is covering areas of "the system of interdependencies", "a flexible latticework of tensions", "a fluctuating balance of power", as well as "patterns created by interacting actors" (Elias, 1978; Burkitt, 1993).

Just as relations and actions of individuals are conceptualized as figurations, so it is possible to think relations between states and their strategies formed for certain purposes as a figuration too. If considering the fact that actors involved in the implementation of certain project constitute a particular figuration, their actions and shifts within the figuration can be explained in the context of the chain of interdependencies. Diversity of actors and their constant reorganization speak about of complex mode of game with several levels and changing patter of power balances. In order to better explain the relationship among actors involved in the energy politics, there is a need to explain who the key actors are. The next part focuses on the concept of structural change and elaborates nature of the game as result of triangular relationship provided by Susan Strange.

2.1.2   Structural power and structural change

The argument of causality and interdependence presented in the concept of figuration can be neatly complemented by Susan Strange's theory on structural power of international political economy and structural change, especially when identifying regional dynamics, directions of interrelationships, type of actors and the nature of competition among the particular actors.

Structural power

Strange provides a very diverse approach to the current interstate relations and presents political economy in a totally new line, wherein the analysis of power stays at the center of her argumentation.[7] She distinguishes between two types of power – relational and structural. Structural power is different from relational power, which has been the subject mainly in traditional theories of international relations. Relational power is the ability of actor A to force an actor B to do something that it will not otherwise do. Structural power "is the ability to shape and determine the structures of the global political economy within which other states, their political institutions, their economic enterprises and their scientists and other professional people have to operate" (Strange 1994).

As Elias, Strange also tried to identify the dynamics of relationship among different actors through the conceptualization of power. But the difference is that she has conceptualized the power by analyzing it within the framework of political economy and finance. Power is considered as an important element in international political economy and determined in terms of structures. For Strange, structural power confers "the power to decide how things shall be

---

7   Indeed, analysis presented by Strange cover broader area of issues of international political economy. However, for the purpose of this paper, only certain arguments have been used for conceptualization of pipeline politics and energy security in the Caspian region.

done, the power to shape frameworks within which states relate to each other, relate to people, or relate to corporate enterprises" (Strange 1994).

The relationship between political and economic is strongly interlinked. A clear distinction between political power and economic power is almost considered as impossible. For Strange, political decisions and economic actions constitute different sides of the medallion, mutually influencing each other (Strange, 1992; Strange, 1995). Besides, "it is impossible to have political power without the power to purchase, to command production, to mobilize capital. And it is impossible to have economic power without the sanction of political authority, without the legal and physical security that can only be supplied by political authority" (Strange, 1994). Hence, within such constellation of interdependency, economical power and political power constrain and enable each other mutually.

Strange also differentiates two levels of structures – primary and secondary – where political and economic are equally presented. Security, production, financial and knowledge are defined as primary structures of power. Energy, transportation networks, trade and welfare are considered as secondary structures and presented as a product of the four primary structures (Strange, 1994). According to Strange, economic or political developments are conditioned by primary structural power (May, 1996).

The central focus in structural theory is given to four primary structures of power. Before analyzing the dynamics relationship among actors, one has to go through the description of these structures. According to Strange, the security structure in political economy is the framework of power created by the provision of security by some human beings for others, those who provide the security acquire a certain kind of power which lets them determine, and perhaps limit, the range of choices available to others" (Strange, 1994). With other words, power in the security structure flows from provision of the security by one group for another. Hence, an actor performing well in terms of security structure will acquire relative advantages in other domains. Balance of power is taken as main base for security analysis. However, it does not get primary position within the whole system. As very few conflicts between actors in the international political economy are pushed as far as the utilization of military force, power in the security structure is not the conditioning structure of international political economy, but instead is only a special case, subject to pressures from the other three structures (Strange, 1994).

The production structure is the sum of all the arrangements determining what is produced, by whom and for whom, by what method and on what terms (Strange, 1994). This is the structure, which creates wealth within the political economy and determines the nature of competition among the actors and the dynamics of state – market relation. The effect of the production structure on the nature of competition will be elucidated more broadly later within the following sub-chapter.

The financial structure, according to Strange, is the key for economic power. Here she underlines the importance of credit that can be created. Strange emphasizes that what is invested in modern economies is not money but credit, and credit can be created – it does not have to be accumulated (May, 1996). In addition, she states:

> "The power to create credit implies the power to allow or deny other people the possibility of spending today and paying back tomorrow, the power to let them exercise purchasing power and thus influence power for production, and also the power to manage or mismanage the currency in which credit is denominated, thus affecting rates of exchange with credit denominated in other currencies" (Strange, 1994)

Therefore, whoever can gain the confidence of others in their ability to create credit will control the economy (Strange, 1994). In fact, the power to create or control the creation of credit reflects the influence over purchasing power and the ability to influence markets for production.

The knowledge structure, which is hard to define, is based on the assumption of what is believed, what is known (and perceived as understood or given), and the channels by which these beliefs, ideas and knowledge are communicated, or confined, making its influence and role hard to assess (Strange, 1994; May, 1996). Power in the knowledge structure lies as much in the capacity to deny knowledge, to exclude others, as in the power to convey knowledge and include others (Strange, 1994). Consent rather coercion constitutes the basis of the power structure, because authority is recognized based on commonly accepted systems of beliefs.

Strange's analysis may imply that changes in the four structures of power are altering the way in which the international political economy is organized. Also, actors may show different levels of performance in all the four structures. On the other hand, these four structures interact with each other within the framework of a system.

In order to explain the power relationship among actors and its outcomes in international political economy, Strange raised the question "Who benefits?"

The answer to this question helps to identify the actors of the structural power, balance of the interests within the structure and the actors' bargaining power. For Strange, it also identified three interconnected aspects of the international political economy that are conditioned by structural power – the continual bargains being struck between authority and market; the ordering or prioritizing of values in any outcome; and the allocation of risk/allocation of benefit. She emphasizes:

> ... it is impossible to arrive at the end result, the ultimate goal of study and analysis of the international political economy without giving explicit or implicit answers to these fundamental questions about how power has been used to shape the political economy and the way in which it distributes costs and benefits, risks and opportunities to social groups, enterprises and organizations within the system (Strange, 1994).

Since politics and economics are interlinked with each other within the structural configuration, the relationship between the two can be analyzed by focusing on the effect of political authority (not only states) on markets and conversely, of markets on those authorities. In fact, not only do political decisions affect market dynamics, also market dynamics also affect political decisions. If economics concerns the allocation of scarce resources, and politics concerns providing public order and/or public goods, then any theory bringing them together must take these different foci into account (May, 1996).

By focusing on the state-market and market-state nexus, Strange argues that it is power that determines the relationship between the two. On the other hand, power distribution among the market players is very crucial within this context. Moreover, it is not only the direct power of authority over markets that matters, but also the indirect effect of authority on the context or surrounding conditions within which the market functions (Strange, 1994). While analyzing the relationship between state and market, it is also necessary to analyze the decision-making process, i.e. why decisions were taken as well as who took them (May, 1996).

While analyzing the process of bargaining the values, risks and benefits should be included into the process. These factors do matter while setting positions and measuring the outcome during the bargaining process. Strange defines four basic values provided through social organization: wealth, security, freedom and justice. In addition, she stresses that different societies may have different objectives and priorities. Hierarchical ordering of the values determines actors' actions according to their priorities. Though all societies are structured on these four values, the combination of these values differs from one structure to another one. Strange stresses that in fact, it is power that determines the nature of the combination (Strange, 1994). For example, some will give the production of wealth in material form the highest priority (Strange, 1994); in this case, the power will be associated with the ability to create the wealth.

In the theory of international relations it is accepted as normal that states should ally themselves with others while remaining competitors, so that the bargaining that takes place between allies is extremely tough about who takes key decisions, how risks are managed and how benefits are shared. Risks and benefits are introduced together with the concept of structural power. Even today, issues related to the perceptions, mitigation, allocation and management of risks constitute an important part of economic and political analysis. The main question in this regard is "how have markets and states created risks, and how have they attempted to mitigate them, or to convert them into costs?" (Strange, 1983). Analysis of risks also covers the discussion of opportunity. An economic approach to this issue is concerned with opportunities for the creation of wealth. Identifying risks helps reveal the balance between state and market.

Structural change

The development of the global economy and structural power are leading to the structural change within the system. In order to explain the change in the global economy Strange developed three arguments. One is the changing nature of the competition among states. With the development of science and technology the nature of the competition between states has altered. If in the past states competed for control over territory and wealth-creating resources – whether natural or man-created – within those territories, now they are increasingly competing for market shares in the world economy (Strange 1992; 1995). Strange also argues that the possession of natural resources are not the most important factor to win market share. Indeed, resource – poor states may win market share by entering into the markets with share in products or services where high value-added output offers better profit margins. (Strange, 1995). Moreover, understanding of power and competition for power have been reformulated as a result of the global change. States are now competing more for the means to create wealth rather than for power over more territory. Where states used to compete for power as a means to wealth, they now compete for wealth as a means to power – but more for the power to maintain internal order and social cohesion than for the power to conduct foreign conquest or to defend them against attack (Stopford, Strange, & Henley, 1991).

The second argument put forward by Strange is the growing role of firms in international political economy. For a long time, state and interrelations between states were the main units of analysis in international relations. Meanwhile, the structural change of international system has led to active involvement of big firms in the global market. Today firms engage not only in local markets, but also seek additional markets abroad in order to get profits necessary to amortize their investments and ability of competition (Strange, 1992). Firms like states start to play a significant role in the determination of the strategic directions for development in the world economy.

Structural change of the international political economy has altered the nature of the game by affecting the actions and moves of all actors, including states and firms. Moreover, the emergence of new forms of global competition among firms has influenced the competition between states. As firms harness the power of new technology to create systems of activity linked directly across borders, so they increasingly concentrate on those territories offering the greatest potential for recovering their investments (Stopford, Strange, & Henley, 1991). Besides, markets stopped to be national and become multinational.

On the background of all these and other changes taking place in the world system, Strange developed her third argument. With the entrance of new players, namely firms, the nature of diplomacy has fundamentally shifted by increasing mutual interdependency between firms and states. In this way the dynamics of the game became more complicated. The cross-border competition among firms and states start to flow in three dimensions – state-to-state, firm-to-firm, and firm-to-state – where governments must now bargain not only with other governments, but also with firms or enterprises, while firms now bargain both with governments and with one another (Strange 1992). Figure 1 illustrates the triangular relationship of new dimensions of actors' relations.

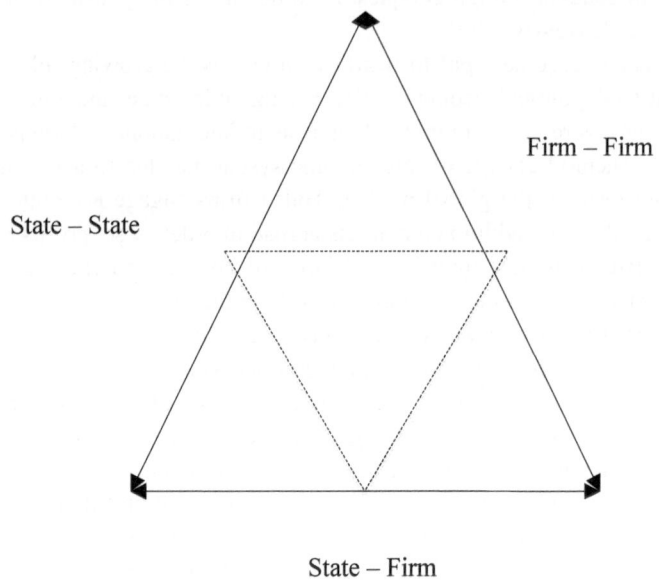

State – Firm

Figure 1: Triads of relationship Source: (Stopford, Strange, & Henley, 1991)

Including the two new dimensions of relationship, it is possible to determine the following directions within the given triangle: the bargaining among states for power and influence, the competition among firms contesting the world market, and the specific bargaining between states and firms for the use or creation of wealth-producing resources (Stopford, Strange, & Henley, 1991). Since all three directions constitute the dimensions of one particular figuration, they become mutually interdependent. The change or decision taken on one side of triangle will affect the other sides as well. As in inter-state diplomacy, in triangular relationships elements of conflict and cooperation exist simultaneously in the bargaining process (Stopford, & Strange 1991). Hence, success or failure in such constellation should be explained through considering the moves taken on all three sides of the triangle.

The current process of bargaining in the international political economy involves not only states, but also national and international companies/firms. Moreover, states are not always able to determine the result of the bargaining, since the power relation can vary. The outcome of the bargain will, however, reflect where the main structural power lies in that relationship (Strange, 1994). Additionally, since power is not static, power balances are tending to change among actors of the game. That is why it is difficult to say which has more power and when– states or firms. According to Strange and Stopford, change in the international political economy points to the fact that states are losing power to pursue independent policies and now must master the new game of triangular bargaining (Stopford, Strange, & Henley, 1991). At the beginning a government may hold power to control and regulate the outcomes when a firm enters the country, but may lose it over time. However, when firm starts operation in the host country, they become more interdependent from each other in different directions. Moreover, changes in the international political economy create new sources of asymmetry relevant for three sides of the triangle. The growth of global competition can be seen as moving the world towards a position where events are conditioned more by an emerging managerial technocracy than by traditional nations of state power (Stopford, Strange, & Henley, 1991).

Since states and firms play together a distinctive role in world economy and politics, it is impossible to note any actions or moves, especially in the energy sector, either with purely economic rationale or political rationale. All actors involved in the particular figuration have to manage multiple agendas. Indeed, within the framework of complex interdependent networks, it is difficult to set clear strategies and maintain explicit policies. That is why complexity leads to greater reliance on implicit policies, or even in extreme cases to policy setting by default (Stopford, Strange, & Henley, 1991).

Compared to firms, states have to manage a series of difficult trade-offs among competing internal and external objectives. They run considerable risk if they fail either to keep some of these agendas separate or to manage the separation in such a way as to appear inconsistent and thus to lose the confidence of prospective investors (ibid). Also, one of the main tasks of states is the ability to identify strong firms as a partner.

Firms are also challenged by the problems caused by growing complexity. As firms have to deal with a limited number of issues in contrast to states, they have better opportunities for maneuvering and command the structure. However, managers at local level may challenge their ability to command from the center, because multinationals are far from the monolithic actors they are often deemed to be (ibid).

Since the understanding of power and power relations among actors of international political economy changes, where firms start playing an important role, economic factors become more decisive when determining foreign policy strategies. A new model of diplomacy, where firms are key decision makers, can be easily observed in the case of energy politics. The following part, which focuses on radical geopolitics draws a more precise picture of how new directions of diplomacy, affects energy politics.

2.1.3   Radical geopolitics and logics of power

The relatively young term "radical geopolitics" is sketched on Arrighi's and Harvey's concepts of two logics of power: territorial and capitalist logics. The concept is developed by Mercille, who reformulated two logics into the geopolitical and geo-economic logics respectively. Both logics intend to determine political and economic drivers behind the foreign policy of the modern capitalist state. In order to be able to understand the ideas put in "radical geopolitics", one has to have a closer look on Harvey's concept on logics of power.

Two logics of power, namely territorial and capitalist logics, are conceptualized by Harvey's theory of capitalist imperialism, defined as 'a dialectical relation between territorial and capitalistic logics of power, where territorial refers to the political activities of state managers and the capitalist to the activities of firms and the processes of capital accumulation (Mercille, 2008). Despite the differences between the two logics, they are tightly interwoven and play a significant role in policy formation. For Harvey it is impossible to think about these two logics separately form each other or reduce one's significance compare to another one (Harvey, 2003). By focusing on the dialectical nature of relation between these logics he emphasizes:

> The relation between these two logics should be seen, therefore, as problematic and often contradictory (that is, dialectical) rather than as functional or one-sided. The dialectical relationship sets the stage for an analysis of capitalist imperialism in terms of the intersection of these two distinct but intertwined logics of power. The difficulty for concrete analyses of actual situations is to keep the two sides of this dialectic simultaneously in motion and not to lapse into either a solely political or a predominantly economic mode of argumentation (Harvey, 2003).

Harvey's distinction of capitalist and territorial power is taken form Arrighi's logics of power. However, his approach enormously differs from Arrighi's. Arrighi conceive them as opposite modes of rule or logics of power.

> Territorialist rulers identify power with the extent and populousness of their domains, and conceive of wealth/capital as a means or a by-product of the pursuit of territorial expansion. Capitalist rulers, in contrast, identify power with the extent of their command over scarce resources and consider territorial acquisitions as a means and a by-product of the accumulation of capital (Arrighi, 1994).

Arrighi defines both logics primarily as forms of state policies and understands it as a mode of rules, whereas Harvey recognizes capitalist and territorial logics in terms of the distinction between the economic and the political. For Harvey, the former is associated with "the molecular processes of capital accumulation in space and time" which occur "through the daily practices of production, trade, commerce, capital flows" and the latter with "the political, diplomatic, and military strategies invoked and used by a state… as it struggles to assert its interests and achieve its goals in the world at large" (Harvey, 2003). With other words, capitalist logic refers to the control of money, assets, the flow and circulation of capital, and territorial logic refers to territorial source of power lying in state organizations (Ashman & Callinicos, 2006).

The radical geopolitics designed on Harvey's conceptualization encompasses a similar approach. Territorial logic introduced later as geopolitical logic is mostly used to explain political decisions and state policies which intend to maintain state's international credibility (Mercille & Alun, 2009). The term of geo-economics logic refers to Harvey's capitalist logic and focuses on the broad political economic aspects of capitalist expansion (Agnew & Corbridge, 1989). In fact, the understanding of the term varies from scholar to scholar. In general the term of geo-economics has been explained in three ways: First, it has been used to refer to politics aimed to maintain, control and exploit natural resources in a certain area (O'Hara & Heffernan, 2006). Second it was conceived as discourse, closely linked with economic imperatives of globalization (Smith, 2002; Sparke, 2002, 2007). Political rationales behind economic movements, such as trade, finance, flows of capital over spaces and across borders have been presented by a third group of scholars (Sidaway, 2005; Mercille, 2008). Considering all three explanations, geo-economics logic within the concept of radical geopolitics is understood as capitalist reasoning of the political decisions and moves. The logic and its impact on state policy is explained through the concept of the "spatial fix," which refers to the physical fixation of capital in places, or to the spatial expansion[8] of capitalist activities (Mercille & Alun, 2009).

Geopolitical and geo-economics logics, in general, reflect political and economic factors, which affect state foreign policy on regional and international levels. The explanation of political factors by referring to geopolitical logic of power can be built on two different spatial scales: the international and the national. Both may give specific impulses in order to influence state policy in different ways. On the national scale geopolitical logic presents, in general, public policy implemented by state at home, where public opinion can be considered as an important element (Mercille, 2008). The international scale is more important in shaping foreign policy, and may be equated in practice with state officials' need to maintain "credibility" (Mercille, 2008).

Starting from 1990s the geo-economics has been considered as more influential than geopolitics while analyzing current interstate relations (see Luttwak, 1993). Since political and economic factors do shape state policies, however, depending on variables and the desired outcome, one may dominate over another one. That is why it will be wrong to neglect and diminish the importance of the political factors compared to economic ones. Although economic and political factors do shape the state policy together and economic forces may predominate in orienting its direction, in fact, capitalist state enjoys a significant degree of relative autonomy in contrast to economic actors and forces (Harman, 1991; Miliband, 1983; Block, 1987; Ashman & Callinicos, 2006; Mercille, 2008). Consequently, the relationship between political and economic should be taken as interdependent. This approach is similar to the Marxist theory of imperialism, where it is argued that political and economic competition have become interwoven in modern capitalism (Callinicos, 2003). Considering the link between these two logics, one may state that players of particular figuration are involved in economic and political competition at the same time.

While conceptualization of the relationship between political and economic factors, the possibility of cooperation and the conflict shouldn't be excluded. Moreover, also Harvey does not exclude the possibility of tension

---

8   The spatial expansion of capitalist activities is closely associated with geo-economics logic, as the "outer" fix resolves (although only temporarily) the tendential over accumulation of capital and labor power that threatens the devaluation of capital. For more details see Harvey, D. (1985). The geopolitics of capitalism. In D. Gregory, & J. Urry (Ed.), *Social relations and spatial structures* (pp. 128-163). London: Macmillan. And Mercille, J., & Alun, J. (2009). Practicing Radical Geopolitics: Logics of Power and the Iranian Nuclear "Crisis". *Annals of the Association of American Geographers, 99* (5), 856 - 862.

between the two logics. Within the framework of radical geopolitics the competition can be analyzed in two ways: broadly it tackles political and economic
competition between states on regional and international levels; narrowly it
focuses on competing interests of diverse groups within a state, namely
between state managers and capitalists. Analyzing competition among various
actors through different dimensions and within the intersection of two forms of
competition – economic and geopolitical – has two merits. First, it includes
both forms of the competition to the nature of interstate rivalry (Harvey, 2003).
Second, it conceptualizes the dialectical relationship between two logics by
considering specific interests of two distinct groups of actors, namely capitalists
and state managers (Ashman & Callinicos, 2006).

Competition among major powers even today remains a crucial feature of
the world's political economy. Focusing on geopolitical aspect of the competition Harvey stated that the dominance of the heartland would assure geopolitical dominance in the Greater Eurasia (Harvey, 2003). The relationship between
state managers and capitalist should be thought as in terms of structural interdependence. The structural interdependence between state and capital leads to
the partnership between state managers and capitalists. However, both groups
of actors may pursue different reproduction strategies and will have competing
interests. Though conflicts may arise between state managers and capitalists,
they may be thought of as interdependent, acting in partnership (Miliband,
1983). Mercille elucidates the interdependency between state managers and
capitalists as following:

> State mangers need the support of economic actors in order to maintain some
> reasonable level of economic growth, as the state's capacity to maintain itself
> through taxation depends on economic activity, and as popular support for a
> government depends in part on the health of the economy. Conversely, capitalists
> (business) need the state for economic regulation, domestically and internationally.
> State agents remain strongly committed to enacting policies that preserve the
> health of a capitalist economy since they are dependent on it for their survival, but
> because state officials and capitalists sometimes examine economic problems from
> different perspective, they may at times favor conflicting policies (Mercille, 2008).

According to Brenner, distinctive groups of actors occupy particular places in
the relations of production and therefore pursue specific strategy in order to
maintain their positions (Brenner, 1986). Hence, the rationale behind capitalists' moves is maintaining and expanding their capital within the dynamic of
competitive accumulation. The risk of failing to do so is the bankruptcy or

absorption by a stronger competitor. In contrast to capitalists, the state managers focus on maintaining the power over their population, extract resources and functionality of state institutions, on the one hand, and balancing power compare to their rival states, on another hand (Ashman & Callinicos, 2006). With other words, the geopolitical logic will be the main driver of state managers decisions/actions and capitalists will follow the geo-economic logic. This discrepancy arises because politicians, in order to achieve their goals in the international arena, require some approbation (coerced or not), from other governments and political actors, in a way, which capitalists do not (Block, 1987; Ashman & Callinicos, 2006). Mercille states the argument and provides the following explanation:

> Indeed, capitalists establish relations in the international system mainly through price incentives and the demand for their products and services. However, politicians interact in the international arena through diplomacy, persuasion and argumentation and thus need to generate at least some consent (which may be coerced) for their policies. One important aspect of this is the need to maintain credibility (Mercille, 2008).

Evidently, due to different rational capitalists and state managers will differently assess their interests at stake. Despite of different motivations, both are mutually dependent, as they constitute players of one particular figuration. Moreover, it is impossible to think about decisions on economic issues without influence of business elites (Block, 1987). In order to guarantee the general conditions for capital accumulation capitalists need state support. On the other hand, state managers need capitalists to be able to maintain the power internally and externally, and to extract the natural resources to the world markets to ensure sustainability of the profit. In fact, in the globalizing world state power highly depends on the capitalist mode of production, size and profitability of the capital based in their territory (Block, 1987; Ashman & Callinicos, 2006). However, one should avoid considering the state as an instrument of capital.

Another difference between these groups of actors is that due to profit pressure, capitalists more often have short-term interests than politicians, whereas those interests may trump the longer-term objective (Miliband, 1983). Nevertheless, despite the differences, both state managers and capitalists are interested in maintaining a credible economic system and a favorable economic climate. Today, energy politics constitutes one of the main pillars of the security policy. Control of the resources and transportation infrastructure provide to major players an ability to regulate and influence the world economy. As long as geopolitical logic and geo-economic logic do not clash firms and state authorities march together in the same direction.

Within the world, where commercial interests matter as political ones, applying and using the concept of geo-economic logic – in lieu of reference to Harvey's capitalist logic – can help to understand the search for the new markets in terms of current pipeline politics in and out the Caspian region.

2.2 Theory and practice

To explore the dynamics of the pipeline politics in the southern gas corridor, this study attempts just to construct a figuration of the regional and non-regional actors. The concept suggests that actors are always connected to one another in a web of interdependencies, whereas fluctuation of the balances of power can restrain or enable a particular actor. The distinction between two logics of power presented in the previous part and the application of the triads of relationship will provide a framework to explore the interdependencies among actors and their interests, as well as to analyze and specify the different outcomes of the decision-making process.

An important as well as challenging question to pose is who the actors are and what kind of decisions (or moves) do let to the shift in the game's dynamics. As it will be seen, pipeline politics is affected by decisions of different actors, including state and non-state. Here, the concept of figuration suggests that a state, in the same manner as relations between states, function like a network of interdependencies which assumes multiple initiatives and introduces, accordingly, the idea of plurality of actors within and between states (Devin, 1995; Isachenko, 2012). The implication of this statement suggests that the concept is not bound to the territorial boundaries of the state, but also differentiates between internal and external spheres. Besides, a plurality of actors emerges as result of the existence the numerous chains within a network of interdependencies, where states and firms interact with each other. Considering the interdependencies between these actors, the study aims to analyze how this interdependency affects decision-making and shifts the current pipeline dynamics.

The combination of some basic conceptual frameworks taken from Elias, Strange, Mercille and Harvey were chosen for a number of reasons. Firstly, the concept of figuration of Elias can be used to explain processes in larger and complex systems, not only the relationship between individuals and society. The idea of network and plurality of actors allows reviewing the processes and interactions among the actors at the different dimensions through the prism of causality. The second important point is that on focusing the dynamic nature of the figuration. The figuration itself is not static and nobody possess absolute power. So the decision (or any move) of the one actor may cause shifts within the whole system, and at the same time determine the actions of the other actors. Considering actors as rational players it is possible to conclude that some actors plan the outcome and each move/decision have logical explanation. Consequently, in the case of the current energy and pipeline politics, one has to consider all direct and indirect factors, the motives of the actors affecting the process and has to analyze the situation from the perspective of interdependency. Elias's concept of figuration assesses the interaction among the actors as a game with multiple dimensions, which changes constantly as the game proceeds and where each stage constitutes the continuation of the previous one.

Strange's concept of structural change will help to understand the dynamics of relationship among actors, including state and non-state, and the nature of the current competition. Focusing on political and economic power, and relation between state and market, it is possible to define what is at the stake for the different actors.

The concept of radical geopolitics is borrowed from Mercille and Harvey. Despite the growing economic ties among various partner states, the geography is still play an important role in determining the directions of the energy politics. The concept of radical geopolitics covers more broad implication of the geography, and helps to analyze the processes from geopolitical and geoeconomic logics. As it is difficult to draw a clear line between political and economic interests of the actors, both have to be taken as intertwined, in order to explain the motives of the key players.

3 Energy Security and Pipeline Politics

3.1 Historical development of the energy security concept

The concept of energy security has gone through transformation throughout history. Each political and financial crisis added a new element to the meaning by extending the approaches. Historical development of the energy security concept initially coincided with the development of the oil industry. Energy security had emerged as an issue of great importance during World War I, when Great Britain switched from coal to oil as the main fuel and became dependent on insecure oil supplies from Persia (Yergin, 2006). The security of energy supply became a question of national security and military success of the industrialized country. At that time, diversification of the sources was the key factor for ensuring energy security[9]. Oil had become the main fuel for transport by 1930 and replaced coal as industry's primary energy source.

During World War II, the significance of the energy resources increased and the control of the oil rich regions composed a central line within the strategic objective of the key conflicting parties (Yergin, 2008). Development of pipeline and supertanker infrastructure for transportation of the oil led the price fall during the late 1950s. Transformation of the world's energy regime after World War II led to the broader usage of the term of energy security, mainly determined by oil prices, demand and supply factors. Oil was an integral part of the world's post-war economic growth trajectory, particularly through the transport sector that currently accounts for 34% of world energy consumption (IEA, 2008a). Oil was relatively abundant and cheap until the oil price shocks or oil crisis of the 1970s.

On October 17 1973, in response to the U.S. decision to resupply the Israeli military forces, members of the Organization of Arab Petroleum Exporting Countries (OAPEC consisting of the Arab members of OPEC) plus Egypt,

---

9   To underline the importance of the diversification, Yergin in his work quoted Churchill, who stated: "Safety and certainty in oil, lie in variety and variety alone." Yergin, D. (2006). Ensuring Energy Security. *Foreign Affairs*, 85 (2), 69-82.

© Springer Fachmedien Wiesbaden GmbH, part of Springer Nature 2018
S. Amirova-Mammadova, *Pipeline Politics and Natural Gas Supply from Azerbaijan to Europe*, Energiepolitik und Klimaschutz. Energy Policy and Climate Protection, https://doi.org/10.1007/978-3-658-21006-9_3

Syria and Tunisia announced a total embargo on all oil deliveries to the USA. Later, the decision on the embargo was extended to Western Europe and Japan, and ended with a price rise by 70% and a production cut by 5%. That crisis and restriction on production lasted till March 1974.

The decision to cut production and raise prices had severe consequences. For the first time in history, oil was used as a weapon (Shaffer, 2009). It underlined the high level of the dependency of the Western countries and its allies on oil (Yergin, 2008). Following the crisis the approach to energy security was reformulated and it was viewed as synonymous of the need to reduce dependence on oil consumption (Martin & Harrje, 2005). Moreover, in a response to the oil embargo of 1973 and the shortfall in global energy supplies, member states of the Organization for Economic Cooperation and Development (OECD) decided to create the International Energy Agency (IEA), in order to help countries co-ordinate a collective response to major disruptions in oil supply through the release of emergency oil stocks to the markets (IEA, 2015).

After the oil crisis of the 1970s, the oil market faced further challenges followed with the Iranian Revolution of 1979 and instability in the Gulf region during the early 1980s. The consequences of the conflict and political instability in the Persian Gulf were severe for the security of supply and oil market that experienced a harsh shortfall of the production and raise of the prices. In order to ensure energy security, the OECD countries decided to increase production in oil producing member countries and expand domestic production capacities. Furthermore, Washington declared the Carter Doctrine, which justified use of the military force to defend US national interests in the Persian Gulf (Jordan, 1982; Brzezinski Z., 1983) stating:

> An attempt by any outside force to gain control of the Persian Gulf Region will be regarded as an assault on the vital interests of the United States of America and such an assault will be repelled by any means necessary, including force (Meiertöns, 2010).

Consequently, ensuring energy security was listed as high priority at the political agenda allowing direct state involvement. After the Persian Gulf War (1990-1991), it was thought that concerns over energy security diminished, the functioning of the oil market would not be subject of political manipulations and supplies would be abundant at prices that would not impede the global economy (Yergin, 2006). However, a decade later, energy markets started feeling shortages influenced by political factors and market dynamics. The global demand for oil increased following economic growth in developing countries, particularly in China and India, and continued economic boom in the western countries in the light of depleting indigenous production (Helm, 2005; Yergin, 2006). Parallel to this, changes within the geopolitical map of the world were causing additional impediments to energy security.

It is worth mentioning that nowadays concerns over energy security are not limited to oil industry and they go much beyond. At present, energy security concepts are changing totally in the new *amplua* which has been determined by the growing dominance of fossil fuels not only in the military sector, but also was influenced by growing consumption in the spheres of industry, heating, electricity production and transportation. Liberalization of the energy markets, the rise of energy demand and formation of the global oil market required a new focus on energy security issues. Moreover, energy security was considered as one of the key factors shaping interstate relations and was reviewed within the political context. Recently, energy security concerns continue to occupy the political agenda as a priority issue, because of broader usage of the fossil fuels in the industry and transportation, and growing dependency from the hydrocarbons.

Starting from Churchill's time and until today oil is vital for the energy and politics. As it was observed during the last two decades, major energy concerns are linked with the security of oil sources, price stability, and demand and supply relations. In fact, the old framework of the energy security was mainly determined by security of oil supplies, and historical experience determined the principles of energy security. The first and most familiar one is diversification of supply. Multiplying supply sources creates an opportunity to respond to and to prevent supply disruptions through alternatives. But diversification is not enough to meet all security concerns. A second principle is resilience, a "security margin" in the energy supply system that provides a buffer against shocks and facilitates recovery after disruptions. As Daniel Yergin describes it, *resilience can come from many factors including sufficient spare production capacity, strategic reserves, backup supplies of equipment, adequate storage capacity along the supply chain, and the stock piling of critical parts for electric power production and distribution, as well as carefully conceived plans for responding to disruptions that may affect large regions* (Yergin, 2006). The third principle emphasizes the importance of integration for stability in the energy market. And the fourth principle is associated with the importance of quality information, which underpins effective market functioning. In addition to the above-mentioned principles, Daniel Yergin completes the concept with two further principles: one is the recognition of the globalization of the energy security system, and another one is acknowledgment of the fact that the entire energy supply chain needs to be protected (Yergin, 2006 & 2008).

The new framework of the energy security requires expansion and inclusion of the new and all chains of the entire energy system. Although oil was the essential part of the energy security concerns, developments in the electricity sector, use of nuclear energy and natural gas also have had a particular impact on the interpretation of the security concept, in general.

A turning point in 1950s and 1960s was the development in the electricity sector, and the use of nuclear energy for power generation. Electricity became fundamental to everything in industrialized world. Emergence of nuclear energy reduced the role of coal. Broader usage of nuclear energy for power generation has been reducing dependency of this sector from oil supply. Besides, nuclear energy was often considered as best option to ensure supply security concerns (NEA, 2012). During the climate change debates there has seen some resurgence in the promotion of nuclear energy as a low-emitter of greenhouse gases relative to other primary fuels for electricity generation. But after the Fukushima crisis the usage of nuclear energy for electricity generation became a security question.

All these and broader usage of Combined Cycle Gas Turbines (CCGT) forced to re-consider the role of the natural gas not only in the power generation system, but also for the energy security concept. Massive infrastructure development for natural gas led to lower prices and the expansion of this particular primary energy source which now accounts for almost a quarter of the world's energy consumption. The consumption of natural gas has doubled since 1980, when its transportation became increasingly available. The world's known natural gas reserves are as large as oil, but geographic[10] and transportation constraints had been hindering its development and usage in previous years. The existing challenges have been removed with the development of technology in pipeline network system and emergence of a global market for liquefied natural gas (LNG).

In fact, natural gas, as oil, plays an integral part of the political game. Today, energy security concerns are also driven by the natural gas supply. More-

---

10   The natural gas resources are concentrated mostly in politically instable and geographically difficult regions. Moreover, former Soviet Union and the Middle East countries hold nearly 75% of known world reserves. Yergin, D. (2011). *The Quest: Energy, Security, and the Remaking of the Modern World.* New York: Penguin Books.

over, the world, particularly the EU member states, experienced the use of natural gas as weapon and political leverage (Shaffer, Energy Politics, 2009) during the Ukraine-Russia crisis (2006, 2009). In the aftermath of the crisis, energy independence became the key issue added to the energy security concept. Of course, it is difficult to speak about independence, when states become interdependent and linked with each other through the gas trade.

3.2 Conceptualization of energy security

On the background of the intensification of trade and interstate relations, understanding energy security also modified. For that reason, energy security has to be reviewed in a much larger context. For different countries the concept has different meaning. Supplier countries often define the concept in terms of security of demand, whereas consumer countries focus on security of supply and price. For major powers, like Russia, security is determined through reasserting state control over the strategic resources in its near abroad and maintaining control over the pipeline transportation and market (Yergin, 2011). Consequently, in a modern world of increasing interdependencies, the further meaning of the energy security will depend much on how countries manage their relations with each other.

The concept of energy security can be differently explained, since it is highly context-dependent. Referring to historical developments and crisis the energy security is mostly defined by four main dimensions: availability, accessibility, affordability and acceptability. These dimensions were also indicated in the World Energy Assessment, which defined energy security as "the availability of energy at all times in various forms, in sufficient quantities and at affordable prices without unacceptable or irreversible impact on the environment" (UNDP, 2004). The Asia Pacific Energy Research Centre in its research paper *"A quest for energy security in the 21st century: resources and constraints"* underlines the importance of the availability of domestic and external fuel reserves, the ability of an economy to acquire supply to meet projected energy demand, the level of an economy's energy resource diversification and energy supplier diversification, accessibility to fuel resources, in terms of the availability of related energy infrastructure and energy transportation infrastructure, geopolitical concerns surrounding resource acquisition (APERC, 2007) as key in energy security analysis.

In recent years, political, economic and social context, in which usage of the term energy security has evolved and placed on the policy agenda, whereas energy security concept is mainly defined through the market perspective. Additionally, addressing environmental consequences of uncontrolled use of fossil fuels also constitute central line in the energy security matters.

Market-centric definitions

Since energy was considered as an integral part of national security and foreign policy, it often was not analyzed from a market perspective. However, markets play a central role in ensuring, enhancing or attaining energy security. Bohi and Toman define energy security as the loss of economic welfare that may occur as a result of a change in the price or availability of energy (Bohi & Toman, 1996). Also, International Energy Agency (IEA) emphasizes that energy security is lined up with adequate supply of energy at a reasonable cost, and it is simply another way of avoiding market distortions (IEA, 1995). Changes within the state and market relations are considered as an important element of energy security:

> Technological developments will affect the choice and cost of future energy systems but the pace and direction of change is highly uncertain. Governments will … have an important role to play in reducing the risk of supply disruptions. Regulatory and market reforms … will also affect supply. Increased competition between different fuels and between different suppliers of the same fuel will tend to narrow the gap between production cost and market prices, reducing monopoly rents, encouraging greater efficiency and lowering the cost of supply (IEA, 2001).

A year later, in 2002, the market definition of the energy security was extended by IEA, and was defined in terms of smoothly functioning international energy markets, thus delivering a secure–adequate, affordable and reliable–supply of energy (IEA, 2002)

Physical availability and price are taken as main indicators affecting security of the energy markets mainly in liberalized markets (Chester, 2009). Indeed, physical availability is described in parallel with the continuity of supply and reliability. Hence, a security of supply risk should be taken as a shortage in energy supply, either a relative shortage, i.e. a mismatch in supply and demand inducing price increases, or a partial or absolute shortage of energy supplies (Scheepers, et.al., 2006).

A more extended definition of energy security can be found in the EC's "Green Paper towards a European Strategy for the Security of Energy Supply":

Energy supply security must be geared to ensuring, for the well-being of its citizens and the proper functioning of the economy, the uninterrupted physical availability of energy products on the market, at a price which is affordable for all consumers (private and industrial), while respecting environmental concerns and looking towards sustainable development ... Security of supply does not seek to maximize energy self-sufficiency or to minimize dependence, but aims to reduce the risks linked to such dependence (European Commission, 2000).

Malfunctioning of the market should be overcome through proper actions and strategies. As Noël emphasizes, the purpose of energy security strategies is to overcome situations when energy markets do not function properly and should be aimed at making markets work (Noël, 2008). Here, the term of risks requires a further attention. The current analysis of energy security is highly linked with the risk management. Egenhofer and Legge stress that in the current situation security of supply becomes a cost-effective risk-management strategy of governments, firms and consumers (Egenhofer & Legge, 2001).

Depending on the time scale risks can be identified as short-term and long-term. Short-term risks are generally associated with supply shortages caused by accidents, terrorist attacks, extreme weather conditions or technical failure of the grid. Long-term risks are analyzed from *economic* and *political* perspective. Supply shortages caused by inability of the upstream country to deliver sufficient quantities of energy, because of market balance change (e.g. unexpected demand growth) can be considered as an economic risk. Political risks are linked with government's policy to suspend deliveries, or a war, or civil war that prevents exports (Egenhofer & Labory, 1998). Inability of the government to develop adequate risk management strategies and regulate the delivery process is also a political risk.

The European Commission has identified the following risks in its Green Paper on security of supply of 2000:

- *Technical risks* include systems failure due to weather, lack of capital investment or generally bad conditions of the energy system.
- *Economic risks* cover mainly imbalances between demand and supply due to a lack of investment or insufficient contracting.
- *Political risks* outline potential government policies to suspend deliveries due to deliberate policies or war or civil strife or as a result of failed regulation, which is referred to as regulatory risk.
- *Environmental risks* describe the potential damage from accidents (oil spills, nuclear accidents) or pollution, including pollution whose effects are less tangible or predictable (e.g. greenhouse gas emissions).

According to Chester the following risks caused by market instabilities, technical failure or physical security threats can be identified:

- risk of interrupted, unavailable supplies;
- risk of insufficient capacity to meet demand;
- risk of unaffordable energy price;
- risk of reliance on unsustainable sources of energy (Chester, 2009).

Assessing security of natural gas supply, Stern defines risks associated with the
sources of gas supplies, the transit of gas supplies and the delivery facilities
through two major dimensions (Stern J. , 2002):

- *short-term* supply availability versus *long-term* adequacy of supply and
  the infrastructure for delivering this supply to markets;
- *operational* security of gas markets, i.e. daily and seasonal stresses and
  strains of extreme weather and other operational problems versus *strategic*
  security, i.e. catastrophic failure of major supply sources and facilities.

In contrast, researchers of the Center for European Policy Studies in the study
"Security of energy supply: a question for policy or the markets?" reviewed
energy security in short, medium and long-term dimensions, stressing the fact
that supply risks may vary over time. The security of supply is defined as a
cost-effective risk-management strategy of governments, firms and consumers.
According to the study, in the new context of the energy security the prime
responsibility for achieving security of supply has moved from government to
all market participants (CEPS, 2011), because market liberalization and growing economic interdependence between all parties have affected states' ability
to react on energy security issues alone. CEPS report defines security of supply
as a cost-effective risk management strategy, which is the collective responsibility of governments, firms and consumers, resting broadly on three pillars. The
first pillar is energy efficiency, which increases the flexibility of the energy
chain and provides an additional margin of security or achieves the same
security margin at a reduced cost. The second pillar is technology development,
which is essential to ensure efficient production and use of energy and to cope
with environmental challenges. The third pillar is supply optimization, by
which we mean diversification by fuel and region and support for the proper
functioning of the market, which should increase the number of market participants and thereby the flexibility and resilience of the system (CEPS, 2011).

There is a clear distinction between short-term and long-term risks related
to the energy security. Short-term concerns are primarily linked with continuity

and reliability of supply, while long-term concerns focus on fuel availability, including network investment, since the energy security will be significantly influenced by the development of the exploration and extraction technologies. Indeed, supply risks can vary depending on fuel type. Long-term risks relating to oil are mainly associated with ensuring sufficient investment to develop and physically deliver the necessary oil to the markets, as well as the ability to manage the political risks associated with supplier countries. In natural gas, however, long-term security of supply relates to investment and political risk. Considerable investment is needed for infrastructural development, especially upstream and in storage (CEPS, 2011). On the producers' side, it is important to feel confident that they will be able to sell the gas on the wholesale markets and, at the same time, to get clear market signals to assess the commerciality of the potential projects.

The main issue of concern is the reliability and continuity of available technological and commercial mechanisms, which convert primary energy sources for end-use by consumers. Long-term risks concern the adequacy of supply to meet demand and the adequacy of infrastructure to deliver supply to markets which will, in turn, depend on levels of investment and contracting, the development of technology and the availability of primary energy sources (Egenhofer, et al., 2004).

Today, supply security is considered to be the main concern of the energy security. As in the case of the EU, the priority of energy security is to minimize the EU's import vulnerability, supply shortfalls and potential supply uncertainty given the high dependence on one single gas supplier (European Commission, 2007). Particularly in Europe, energy security is strongly linked with the natural gas supply security, which plays a significant role in power generation. Without doubt, it would be wrong to generalize the security concerns taking into account the differences between the energy markets. Energy security in the gas market differs from energy security concerns in the oil market. Therefore, security concerns of concrete markets should be reviewed separately.

3.2.1 Energy security from consumer and supplier perspective

Energy security from consumer perspective is more neatly studied compared to supplier perspective. The former focuses on supply security and the latter on demand security. There are broadly developed theoretical and methodological frameworks to analyze energy supply security, whereas elements of the demand security is less explored. Indeed, security of demand can be as important as supply security, since both are interdependent and compose two different sides of the medallion. In order to understand the correlation between supply and demand securities, different aspects of the supply security are explored in this part of the study.

As mentioned previously, security concerns of the energy markets may vary depending on the fossil fuels. Key elements of the supply security of the natural gas presented in this part are borrowed from Jonathan Stern's studies that are considered as a fundamental work in this field. According to Stern, natural gas supply security covers threats of supply and price disruptions arising from risks associated with the sources of gas supplies, the transit of gas supplies and the facilities through which gas is delivered (Stern, 2002). Referring to traditional security framework the scholar addresses essential questions such as reserve production ratio, long-term contracts and investment commitments, import and transit dependence, commercial and political risks, and examine risks related to the source, the transit and the facility.

The first part of the traditional approach to security of supply includes reserves and reserves to production ratio, which determine the level of self-sufficiency of a country during a certain period of time. The second part is based on supply-demand balance and estimation of the existing adequate supply arrangements to meet expected demand. As a result of dominant position of the long-term contracts in the European markets, major gas companies experience short-term surplus, thus leading to prevention of new entrants reaching the market and delays of new gas-fired power stations.

Although long-term contracts cause short-term surplus, the question is whether in the longer term a new large-scale supply can be obtained for the future. Considering the long-term contracts as one of the major challenges to the supply security, Stern underlines importance of changing in the form of the long-term contracts and suggests changes in the length of the contracts, take-or-pay obligations and price indexation[11]. However, these changes are linked with

---

11   According to Stern the changes should be done in the following ways: Contract length is shortening, such that hence forth long-term will be more likely to mean 8–15 years, rather than 15–25 years. Take-or-pay obligations – traditionally 80–90% of the annual contract quantity – may be reduced, perhaps to 50–60%. Oil-linked pricing and indexation is changing in favour of floating indexation to a product with immediate relevance to the customer, e.g. a gas or electricity spot or futures price in a relevant location. Such indexation guarantees the buyer that prices will remain competitive with other gas supplies. The emergence of a spot

another more important security issue for both consumer and supplier. The absence of traditional long-term contracts and take-or-pay obligations may undermine the willingness of the partners to invest and support new multi-billion-dollar infrastructure projects in the remote areas (Stern J. , 2002). For the producer countries that try to enter the market, especially from the Caspian region, lack of investment can jeopardize the realization of the new pipeline projects. Consequently, multi-billion-dollar projects increase the level of inter-dependency among suppliers and consumers.

Dependency on imports is reviewed as the third part within the traditional framework, where source dependence, transit dependence and facility depen-dence are classified as major risks arising from import dependence. Source dependency in the best case is balanced through the diversification. During the poor diversification and long pipeline transportation, transit risks present other obstacle for the importers of the natural gas from the remote areas. Because, each border crossed adds an additional layer of security risk with the potential for conflict within these transit countries, and between the latter and the supp-lying country (Stern, 2002).

According to IEA, facility dependency and risks associated with it can be considered relevant for all members along the supply chain. As it is emphasized the greatest risk of prolonged interruption comes from the destruction of a ma-jor production or processing facility or a deep-water pipeline whose replace-ment might take many months to build (IEA, 2000; Stern, 2002).

In reality, while looking at the risks from the supplier approach the last two risks constitute potential risks for them as well. Nevertheless, source de-pendence is can be replaced by the market dependence for the supplier. Asses-sment of the supply disruptions reasons shows that throughout the history tran-sit risks were the real security challenge for both downstream and upstream countries connected via pipelines. Hence both have to seek ways of mitigating potential transit risks.

Despite the fact that technical risks can cause damages to the natural gas trade, political and commercial interest of certain players can lead to more serious problems and end up with supply cuts. Considering the fact that hydro-carbon reserves close to traditional energy markets are being depleted and the huge volumes of production moved to the land-lock areas and the need for

---

market assures buyers that they will be able to on-sell volumes surplus to their requirements, rendering take-or-pay obligations much less onerous. For more details see: Stern, J. (2002). The Security of European Natural Gas Supplies.

more cross-border pipelines has increased. Despite the given existence of the LNG transportation, most of the natural gas delivery still will be implemented via pipelines form the land-locked areas. Accordingly, in the case of cross-border pipeline transportation risks associated with the natural gas supply increase, as a result of inflexibility of the route, source and market. In other words, specific nature of the gas trade and market does not allow changing the destination of the pipeline in one day and it becomes the subject of the political manipulation. Moreover, this situation can be used as a commercial or political weapon.

3.2.2   Limits of state's involvement in the liberalized energy markets

Throughout the last three decades, most developed countries started the process of privatization, restructuring economic system and developing deregulation programs in several industrial sectors as a consequence of globalization. For many years states used several mechanisms to regulate the structural behavior and performance of the several markets for goods and services. Sectors such as airlines industry, energy, telecommunication, post services, railroads and etc. have been identified as strategically important and were regulated by State-owned companies or natural monopolies. During the post-world war period politicians were arguing that, some industrial sectors including energy should be under the government's control, since markets fail to provide certain level of security.

Indeed, changing economic environment has affected market dynamics in a way that markets became more independent and self-regulated bodies. It led to the decline of state power in existing state-market constellation and ease of the state's influence. Supporters of the liberalization process claim that markets function much better with less state control. The successful liberalization policy of the capital and financial market has encouraged the liberalization of the other strategically important industrial sectors, including energy. Compared to other sectors, liberalization of energy markets, namely electricity and natural gas have been challenged by various factors, including economic and political dimensions.

Energy stays at the heart of economic growth and production. From the beginning energy has been treated as special and states have always underlined its importance for their security. The basic industries in every modern economy

– steel, chemicals, engineering – all need large inputs of energy, whether this comes from oil, coal, gas or nuclear power. Since the main industries are highly dependent from the usage of energy, disruption of the supply of power will almost lead to the standstill (Strange, 1994). Until the 1980s, it was a conventional wisdom of the post-war years that markets are hopelessly inadequate in providing appropriate energy supplies. State-owned companies were deemed to be so natural that they were made statutory monopolies and it was assumed that regulation was inevitable (Helm, 2003). Moreover, pursued energy policies have been designed in a way to maximize state income. Indeed, the importance of energy was linked with the assumption that state has to control its production and distribution.

With the start of the energy market liberalization, at the end of the twentieth century, the nature of state's involvement in the energy system has changed. By increasing the number of the market participants liberalization has reduced the government's scope for intervention and altered the policy instruments at its disposal. In fully competitive markets the prime responsibility for achieving security of supply shifted from governments to all market participants and companies begin to play a significant role in defining market strategies and ensuring energy security.

Under the process of liberalization state's involvement in the energy markets has been constrained in two ways. First, the self-regulating nature of the markets has left little space for the governments to intervene directly and influence market dynamics. Instruments used before by governments to regulate the markets needed to be transformed. Second, in order to reduce monopolistic behavior of the states in the energy markets, certain legal framework has been developed. Within the new context the main task of state is to develop cost-effective risk-management strategy to prevent energy supply disruptions.

This part elaborates what kind of role a state plays in liberalized energy markets and illustrates different stages of the liberalization process and shows in which areas state's involvement is required. Besides, there is drawn a link between supply security and liberalization. Since, states were using supply security concerns as an arguments to intervene into the market dynamics, it is worth to analyze how liberalization affects supply security.

3.2.2.1   State and market relations

There can be three different, but related forms explaining the link between
states and markets. First, states are integrated into markets through the hierarchical structure of international trade and production (Wallerstein, 1989). Second, states compete with one another to attract mobile capital and 'core' states
struggle over the power to organize the global economy (Arrighi G. , 1994). If
previously states were competing for territories, now they compete for the market share and national companies act as governments agents. Third, state-
market interaction diffuses under conditions of liberalism and globalization.
According to Meyer, globalization certainly poses new problems for states, but
it also strengthens the world cultural principle that nation-states are the primary
actors charged with identifying and managing those problems on behalf of their
societies (Meyer, Boli, Thomas, & Ramirez, 1997).

Globalization of the economy by influencing the interplay between states
and markets has changed the traditional structure and affected the responsibilities of the market players. The old school of comparative political economy
was analyzing states and markets as self-containing separate entities battling in
a zero-sum game for their share of a finite economic space (Block, 1994). For a
long time, state-market relations was taking place within the national economic
space. In fact, terms and rules determined by globalization moved products and
process from national markets to international. National or state-owned companies entered in foreign markets become engaged in a political juggling act
(Strange, 1995). Indeed, national companies are profit oriented entities pursuing their commercial interests as other market players. If at the beginning the
national companies were acting as agents of their governments, by time they
become independent players of the world economy. Now governments must
bargain not only with other governments, but also with firms or enterprises,
while firms now bargain both with governments and with one another (Strange,
1992). With the entrance of the national companies and large firms states have
been forced to share their power with other market players.

Jaskow has defined the globalization-led weakening of the state control as
a deregulation. However, deregulation does not mean the exclusion of government institutions from the markets, per se. It is a complex process that involves
the relaxation of government controls over prices and entry in some industries,
industry restructuring and privatization, the introduction of new regulatory
mechanisms in industry segments and the adoption of market-based mecha-

nisms. Despite regulatory reforms, there are no markets in the world left without regulation of state related institutions. Markets in all modern developed market economies operate within a basic set of governance institutions or what Williamson has called the basic institutions of capitalism (Williamson, 1975). In reality, government exercises its control through common law institutions – property rights, liability rules, contract laws – and market institutions created by statute, such as corporate law, including the framework for creating limited liability corporations, antitrust laws, bankruptcy laws, employment laws, environmental laws, etc. (Joskow, 2010).

3.2.2.2   Energy market liberalization

Despite the ease of the state power in energy market, it still plays a crucial role. In fact, the liberalization has strengthened state's role as a regulatory body. If state is not involved in market directly, it acts as the main player of the liberalization process.

The role of the state in market liberalization process and supply security is threefold: developing competitive environment; granting freedom to market actors through liberalization reforms; preventing; development of cost-effective risk-management strategy. In the liberalization process of the energy markets state has to deal with the following issues: the redesign of the horizontal and vertical structure of the industry, defining competitive market segments, guaranteeing a non-discriminatory access to the network infrastructures, the development of the regulatory institutions.

a) Redesign of vertical and horizontal structure

The liberalization process has started mainly in competitive segments of the energy markets, where monopoly elements persist and networks (transport/transmission and distribution) are key elements[12]. Before the liberalization reforms began in most developed countries, the energy markets were less competitive and used to be managed by national companies. Big companies by maintaining dominant market position often enjoyed economies of scope and

---

12   Competitive segments of the natural gas markets are downstream and midstream supply. The upstream competition in the gas market is a complicated issue, since the major gas fields are concentrated in few countries.

scale, where conglomerate, horizontal and vertical integration have been consi-dered as a norm. In fact, different sectors of the energy markets require institu-tional transformation of regulated monopolies and unregulated competitive segments of the energy industry. State-owned companies are often active in vertical and horizontal integrated structure. For a market to function indepen-dently and fully it is necessary to implement vertical and horizontal restruc-turing (separation), redesign competitive segments and to develop a compatible regulatory framework.

Vertical separation is characterized as an important factor in fostering competition. However, it is a very complicated issue. There are two conflicting positions. Supporters of the vertical integration are describing vertical integra-tion's advantages for the gas sector. The one of the main challenges of the ener-gy sector is the burden of *long-term investment in the upstream phase* (gas contracts; infrastructures), which is supposed to require the need to minimize the uncertainty to sell the gas purchased in international markets (Polo & Scar-pa, 2003). From the beginning most of the gas contracts were designed on take-or-pay principle. Such contracts often are signed between a State-owned com-pany (producer) and a large buyer (not necessary to be a State/national com-pany), who imports and resells it at a wholesale. Since the price and quantity of the natural gas are agreed by producer and importer within the TOP contracts, both sides bear the price and quantity risks respectfully. It is often claimed that the vertical integration naturally can prevent the producer company, which sink the huge investment in extraction and transportation of the natural gas, from the risks caused by market dynamics.

The supporters of the liberalization process argue that the vertical integra-tion is not necessary to fulfill TOP obligations. Despite the constraints posed by TOP contracts, the importer and the seller of the natural gas can be different companies. In order to prevent dominance of a single economic entity in the national markets, an import contract can be divided into several subcontracts. An independent transport network, which is integrated neither with upstream nor downstream companies, will be able to secure that the gas can reach the final destination.

In contrast to vertical integration, it is difficult to prevent monopolistic be-havior of one or group of companies within the horizontal integration. Since the gas fields are situated in few countries and transported mostly with pipelines or LNG, economy of scale is less specific for this sector. The main task of the state is to achieve diversification in the national market.

Apart from horizontal and vertical integration, market liberalization also considers restructuring conglomerates. There are several firms operating in both gas and electricity sectors at the same time and providing multi-services. Companies pursuing such a policy have the following objectives: a) save on costs; b) provide customers with an integrated set of services (one-stop-shop); c) use their strong market position in one sector to induce "captive" customers to buy a bundle of services (Polo & Scarpa, 2003). However, conglomerates should not be seen as a challenge, if the sector is sufficiently open to competition. The first regulatory principle that should apply in such cases where regulated firms are active in competitive markets as well is the separation of accounts, which allows avoiding cross subsidies (Polo & Scarpa, 2003). But on the other hand, the risk occurs, when regulated tariffs do not reflect solely the cost of the regulated segment (ibid). Moreover, company with dominant position can use its market power as leverage to hinder the market opening to competition. The scenario can be different if there would be more companies competing with each-other, and consumers will end up benefiting from the cost savings because of economies of scope.

At this stage of the energy market liberalization, it will be wrong to state that big companies should not be integrated with each other. Indeed, the integration between upstream and downstream companies will consequently guarantee that the product, which is gas in this case, is sold in final market and vertical integration *per se* does not cause problem for competition[13]. It is clear that the liberalization followed by vertical and horizontal restructuring will increase the number of market participants and complicate the coordination. Hence, lack of management can hinder the competition and vice-versa. That is why this stage requires state's direct intervention and regulation. As shown here, liberalization may limit state's involvement on one hand, but also increases its role as a regulatory body on the other hand.

b) Network access and network development

The access to the network and distribution infrastructure is the key element in the energy market liberalization process and Third Party Access principle adopted by the European Union constitutes its milestone. However, the first crucial step is the redesign of the industry structure, aimed at strengthening the market competition by eliminating monopolistic behavior of the network owner in the

---

13 The main precondition in this case is that upstream company does not play a dominant role.

final market. Since access to the network increases competition in the retail supply markets, the participation of the network owner in the final market may challenge the competition development.

Indeed, in a non-regulated access regime the network owner is interested to get high market share and preserve the incumbent profits. If it has no direct involvement in the downstream market, it will be less interested in hindering perspective of the liberalization process. In this case the revenues will depend on access tariffs and network owner will be neutral towards the new entrants. However, if one company is active in different segments of the supply chain, it will be much more interested in enhancing its market power through particular operations. According to Polo and Scarpa, the allocation of transmission rights must be separated from the transactions between upstream and downstream firms. A stricter regulatory regime is required to avoid monopolistic behavior.

The next essential concern related to the TPA is the appropriate level access charges and price setting issue. It should be based on non-discriminatory and cost-reflective principle. When the network owner does not participate in the other markets, this condition ensures that all the firms pay the same access terms, with no undue advantage of some competitors, and that the access price reflects the underlying cost conditions, with no double marginalization effect (Polo & Scarpa, 2003). In a liberalized market nature of the access charges may have variable and fixed components. Indeed the firms in the competitive segments might change their clients and suppliers according to the price movements. Since any trade requires the access to the network, the access tariffs should be sufficiently flexible to allow to change clients or suppliers (and therefore the geographical path of delivery) without paying each time an additional burden. A short time span and the distinction between entry and exit access charges can minimize the transaction costs, still preserving a cost reflective tariff structure (Polo & Scarpa, 2003).

The tariff structure also includes some incentives, namely fixed components, to allow network owner to maintain and develop the transmission infrastructure. However, the regulatory process is challenged by certain factors. That is why it is advocated to keep state control over transmission system under directly managed TPA principle. Development of non-discriminatory access conditions to the network infrastructure is a necessary, but not sufficient condition for the creation of a competitive market. Governments are responsible for development of the competitive environment through the legislation.

3.2.2.3   Market liberalization and supply security

Security of the energy supply is in most cases analyzed from geopolitical perspective and less attention is given on how the changing economic environment affects the supply process. Energy market liberalization and growing international economic interdependence have affected governments' ability to react to security-of-supply challenges (Egenhofer & Legge, 2001). If in the past energy supply security was considered, primarily as the state's responsibility, now it has become a shared responsibility among all the market participants, including the companies. Liberalization policies have alerted the mechanisms and policy instruments used to regulate the markets. As practice shows, fully competitive markets significantly reduce the scope of governments' intervention and at the same time, minimize the supply disruptions.

The link between supply security and market liberalization is complex. Indeed, the flexibility of the energy system may increase with entrance of the new market players as a result of liberalization process. However, the liberalization itself does not eliminate all the risks associated with energy supply, but all market actors share responsibility for the risk management. As a result, security of supply has become a common responsibility shared among firms, governments and, sometimes, individual consumers, with the primary responsibility resting on firms, including both supply companies and large industrial customers (Egenhofer & Legge, 2001).

Growing natural gas demand, rapid increases of energy prices, unexpected supply disruptions and natural gas shortages during cold winter cause significant challenges for the supply security. For a better understanding of the link between liberalization process and supply security we have to clarify what does supply security mean at the liberalized energy market, namely in gas market. Hence supply security is aimed at preventing risks associated with supply disruption. Risk can be defined as short-term and long-term and identified according to the origin: political, economic, technical and environmental.

One of the key indicators of the supply security in competitive markets is price. Change in energy prices affect economic growth and the competitiveness of other related industries. In competitive markets, price plays the role of balancing mechanism for demand and supply. If demand and supply are not in balance, prices change to provide market signals to close the gap (Egenhofer & Legge, 2001). Price being a crucial indicator of the security of supply is considered a measure of economic impacts, reflecting scarcity and depletion of energy resources (Kruyt, Vuuren, Vries, & Groenenberg, 2009). On the other hand, price volatility can be seen as a proof of the markets' functionality. However, price volatility can also be considered as a risk. Sudden price changes and high market prices are affecting economic growth and competitiveness of the related industry. Low-income group of consumers mainly suffer from price volatility, since in short run demand elasticity of the energy is very low and consumer cannot shift to substitutes when price rises (Egenhofer & Legge, 2001).

Moreover, the concept of the supply security can be analyzed in two dimensions: operating reliability in short-term and resource adequacy in long-term. The first dimension of supply security refers to the capability of the system to balance supply and demand in a short-run using existing physical infrastructure. It includes natural gas production, capacity of natural gas storage and transmission network, LNG infrastructure and distribution facilities. The second dimension refers to long-term investments in exploitation of the new fields, development of the pipeline systems, LNG import terminal capacities, storages and etc. Moreover, operating reliability and resource adequacy are interlinked. Operating natural gas system reliably is a more challenging and costly when efficient investments in supply resources have not been forthcoming (Joskow, 2007).

In fact, liberalization does not solve the problems associated with supply security. However, the responsibilities for supply security are much more effectively distributed in fully competitive markets. Increasing the number of market players offers more competitive price in the market and eliminate monopolistic behaviors of certain companies. In liberalized energy markets the balance of responsibility for security of supply has been shifted towards a shared responsibility between the governments, companies and consumers. Changes in the structure of the energy market as a result of the liberalization have shifted market power and the roles of the parties involved. If previously the responsibility for the security of supply laid on market power, in most cases state-owned company, now all players share market responsibility. In competitive markets consumers can be charged for the costs of the increased security of supply. That is why supply security can be seen as cost-effective risk-management strategy of the governments, companies and consumers. However, the instruments used by these parties to mitigate the supply risks differ from each other. The governmental intervention through regulations is required when other parties are unable to solve the problem.

Although the globalization has posed certain limits on state-market inter-connection, states are responsible for shaping the market structure and strategies, guaranteeing their functionality and creating fair environment for the new market entrants. Indeed, the liberalization has transformed the role of states in the market, instead of limiting it.

3.3 Pipeline politics and transportation

Throughout history the significance of the pipeline has changed in terms of power and politics. In the long past, when pipelines were transporting only a small fraction of the oil, it was traditionally considered as an issue with a tertiary economic concern and with little relevance to strategic behavior (Mearsheimer, 2001; Stulberg, 2012). With the rise of the amount of the traded oil and gas via the pipelines, the extra-market value of the infrastructure also increased. A pipeline not only connects supplier with the market, it also lead to dependency between the states along the connection line and can thus be exploited for strategic purpose.

Realists make an accent on the conflict-prone nature of the transit pipelines and consider them as a tool of power politics among the actors. Indeed, pipelines may be a part of resource nationalism politics of the states. The fact that more than 75 percent of global oil and natural gas reserves and that cross-border energy transit falls increasingly under national authority speaks for this impulse to control critical resource endowments (Gilpin, 1983; Hirschman, 1980).

States with preponderant power are presumed to be especially inclined to struggle over pipelines control in their strategic orbit, given the sunk costs and asset-specificity of transit infrastructure (Frieden, 1994). Control of pipeline routes gives power of influence to its owner in strategic regions. That is why, regional suppliers and foreign investors are prone to engage in costly competition to control access to disputed fields and critical transit chokepoints in an effort to obtain influence over important regions and reduce the vulnerability of supply lines (Stulberg, 2012).

On the other hand, the growing energy demand and market share of the energy resources force to see the pipeline politics from another angle. Hence, parties involved in pipeline politics also compete for the market shares. Scholars claiming that market competition will devolve inevitably to 'resource wars', relegating pipelines to service non-commercial foreign policy aims (Klare, 2002; Duffield, 2008).

Neoliberals mostly explain energy transit as a trade and have this approach. Alternatively, proponents of the neo-liberalism provide different perspective of pipeline politics. Since pipeline creates energy interdependence and provides mutual gains, neoliberals claim that it can strengthen cooperation between parties along the infrastructure. Disruption of the energy transit as in the case of the trade disruption will have negative effects on state at the macroeconomic level considering the fact of slow recovery and the costs of disrupting. Hence, economic interdependence radically constrains states in a positive way (Mansfield & Pollins, 2003).

In the framework of the energy security in this case, supply stability determined by the ability to deter, mitigate and contain potential threats to the consistency of the delivery considered as a key factor. Market mechanisms, supply diversification, technological innovation and the availability of strategic reserves constitute 'shock absorbers' to attenuate the negative effects of price volatility (Stulberg, 2012). In this case, pipelines not only render the costs of conflict prohibitive in terms of disrupted supplies, but also provide instruments to soothe otherwise conflict-prone relationship among different parties (Fettweis, 2009). On the other hand, mutual economic benefits driven from energy interdependence and economic interests will deter partner states from deleterious economic effects of a breakdown of the energy trade (Gelpi & Grieco, 2008). In this case, democratic political institutions are seen as critical intervening variables, where increased economic dependence will reduce the trade disruption (Russett & Oneal, 2001).

There are studies proving that conditional and transformative strategies reinforce cooperation between neighboring and partner states (Stern, 2005; Kahler & Kastner, 2006). Globalization of trade and financial markets saps the capacities of states to advance pipeline projects on their own and it increases the availability of foreign direct investment, affording partners more opportunity to communicate true levels of resolve to realize mutual interests in energy transit (Stulberg, 2012). Since states are dependent on private sector, their interests match on maximizing netback values and returns on investment (Jentleson, 1986).

With the emergence of the integrated global oil market in the late 1960s vulnerability of major consumer states have decreased. This can be explained by various factors, but, the main factor was the elimination of asymmetric dependency. Suppliers have become dependent on export earnings and upholding reputations for reliability on the background of shrinking market shares. Even in a tight international oil market or in the natural gas sector, where energy is not traded widely as a fungible commodity, there are strong counterbalancing pressures to import vulnerabilities that render interests interdependent among suppliers and customers (Stulberg, 2012). Consequently, these changes in the global energy market lead to the decrease of the cases with supply disruptions, which had serious damages on the economy of partner states (Shaffer, Energy Politics, 2009).

3.3.1 Transit challenges

In general, oil and gas pipelines have been regarded as 'steel umbilical cords' of dependence, which can be disrupted for commercial and strategic gains (E-bel, 1997). Hence, it is possible to argue that some pipeline projects are politically motivated and constitute elements of various gaming strategies. The large number of routes explains why some projects are ignoring the basic economics of pipelines, which require a critical mass of throughput to be in place before the project can seriously be considered.

The pipeline politics in Eurasia cause scholarly debate over the structural dimension of cross-border energy transit. There are several opinions regarding the nature of the pipeline politics. Neo-realists argue that the pipelines serve as instruments of competitive resource nationalism. On the other hand, proponents of liberalism and Marxism consider pipelines as conduits for constraining opportunism, strengthening interdependence and transforming interests in regional cooperation. Indeed, both review the pipeline politics as a part of the national energy security, where 'the lands between' Russian/ Caspian suppliers and markets in Europe and Asia mere pawns in the global quest for energy security (Stulberg, 2012).

Export pipelines constitute physical-commercial ventures for moving oil and gas, which are subject to economies of scale, long lifecycles, large upfront investment, inflexibility, natural monopolies and the tyranny of distance in the case of natural gas (Stulberg, 2012). There is a widespread assumption that pipelines are highly vulnerable pieces of energy infrastructure. They cross long distance and therefore every mile cannot be guarded all the time. As the experience shows the trans-border pipelines are more vulnerable to the disruptions and cause problems for consumers and producers. Being fixed infrastructure prone to market failure, the commercial value of a pipeline is directly affected by the dedicated upstream supply, price of throughput, availability of alternative supply options and state intervention (ESMAP, 2003). One of the major problems associated with the pipeline transportation is *fragmentation of jurisdiction*. Since trans-border pipeline passes through territories of various states, more than three stakeholders are involved in the process of pipeline construction, operation and rent sharing. These multiple stakeholders are left to their own devices to resolve conflicts of interests, protect vulnerable infrastructure, reconcile different national legal regimes and norms, and locate mutually rewarding outcomes for the reliable delivery of strategically important throughput (Shaffer, Energy Politics, 2009). Moreover, there is lack of regulations and certain gaps within the legal framework. At the same time the mechanisms of sharing profit and rent are not well developed. These three factors – increasingly remote and land-locked oil and gas reserves, growing gas demand and fragmented jurisdiction – have increased the importance of transit pipeline issues in energy politics (ESMAP, 2003).

Certainly, cross-border pipelines are prone to get in trouble in case of political or economic crisis. There are different levels of interdependencies between the parties along the pipeline. Hence, in case of disagreements, it may lead to supply disruption, if parties fail to negotiate and sustain the terms for construction and operation of pipelines. In most cases, pipeline politics refers to the unilateral and arbitrary disruption or renegotiation of the terms of supply, transit, off-take and delivery (Stulberg, 2012).

3.3.2   Limits to obsolescing bargains

Another challenge linked with pipeline politics is rooted in 'obsolescing bargains' of transit agreements between host governments and oil/gas companies. This concept explains a different power relation within the dynamics of transit pipelines by underling the fact that "almost from the moment that signature dried on the document, powerful forces go to work that renders the agreement obsolete in the eyes of the host government (Vernon, 1971). Consequently, in the energy trade, as long as host government depends on foreign investments, it tends to favor the interests of the multinational enterprises. However, the leverage shifts in the favor of the host government over time as the investments are fixed in infrastructure. It gives host governments 'power' to arbitrarily alter the original bargains and to try unilaterally secure a greater share of rent (Stevens, Transit Troubles: Pipelines as a source of conflict, 2009). In this case obsolescing takes the form of renegotiation of transit terms, change in payment procedure and etc. Hence, transit countries tend to be more disruptive and cause the supply interruption in rather than suppliers and consumers with the major shares (Shaffer, Energy Politics, 2009).

Since pipelines attract very large economies of scale and therefore tend to be very large, capital-investment projects, they become vulnerable to the obsolescing bargain because of their structure and physical inflexibility (McLellan, 1992). Transit states tended to renegotiate transit fees, because they want to maximize their shares of value. The actions undertaken by transit country can be threefold. First, the problems get compounded because owners of cross-border pipelines do not always own throughput and the profitability of pipelines derives from near full capacity operation, especially in the case of transportation of the natural gas (Stevens, 2009; Stulberg, 2012). The second is, if the transit country is an active partner in the pipeline, contributing capital and bearing risk, then it is possible to determine what a reasonable rate of return on the transit country's investment might be – i.e. its 'normal profit', although this in itself can be extremely controversial (Penrose, Joffe, & Stevens, 1992). The last logic is linked with the nature of the pipelines. Considering the fact that pipelines are natural monopolies, in order to maximize own profit from the rent, transit states will exploit this monopoly position of supplier by seeking higher returns on its investment given that its sovereign status protects it from antitrust action (Stevens, Transit Troubles: Pipelines as a source of conflict, 2009).

Geography also affects the disruptive behavior of the transit country. Since geography may dictate the transportation route and if there is only one possible transit route, a transit country will have extensive leverage to negotiate the transit agreement in its favor.

Indeed, such behavior of the transit states can be mitigated under certain conditions. Traditional obsolescing bargains will lose its force as host countries become dependent more and more on the intermingling of foreign and domestic investment, seek to move up the value chain with the acquisition of advanced technologies and production assets, and contend with capital and financial mobility in the global economy (Woodhouse, 2006). Transit country will be less prone to disruption, when the government is concerned about country's international reputation, has strong interstate relations with downstream countries, competes with rival pipelines and relies on off-takes from the pipeline. As an active partner in the pipeline project – contributing capital, bearing risk and receiving valuable compensation – a transit state should have less incentive to renegotiate arbitrarily the original terms of delivery (Omonbude, 2007). That is why it is possible to argue that pipeline politics is neither unexpected nor intrinsically unmanageable.

3.3.3   Credibility in energy transit

Considering the risks associated with the pipelines, the key question here should be answered is: why and under which conditions would rational actors seek to engage in costly pipeline politics?

An important part of the answer rests with the bargaining context, where the incentives to forge pipeline agreements are inextricably linked to implementation. This predicament is tantamount to an international commitment problem, whereby the actors – exporters, pipeline owners/operators, transit states and downstream customers – may be unable to commit themselves to follow through on agreements pertaining to construction, operation, pricing and dispute settlement, and may even have incentive to renege on them (Fearon, 1997; Powell, 2006; Stulberg, 2012). However, the challenge of the negotiation process is rooted in asymmetric distribution of information and the level of trust among parties. Trust ensures credibility of the partner and cooperation (Kydd, 2005).

Credibility plays an important role in the planning and implementation of the large pipeline projects. Lack of credibility may question the realization of the pipeline project. Hence, each actor must convince the others that it will continue to cooperate and fulfill its commitments driven from the agreement after construction of the pipeline (Fearon, 1997; Martin, 2005). Similarly, it is possible to impose outside constraints on future shirking. Actors that experience high domestic audience costs from breaking commitments (loss of significant off-take), surrender some sort of bond to a third party by revoking a supply/transit/purchase commitment, or derive legitimacy from an international reputation for trustworthiness, should be poised to go beyond 'cheap talk' (Stulberg, 2012). By ceding sovereign decision-making authority to a third-party dispute resolution mechanism or by 'tying their hands' (via public commitments), states may raise the expected political costs of disruption to make their commitments more credible at the commencement of bargaining (Fearon, 1997; Martin, 2005).

Yet, conventional mitigating mechanism cannot be so easily applied to cross-border energy transit. States involved in such pipeline projects assume that the parties will assign common values and maintain equal capacities to formulate and uphold commitments driven from the agreement. However, these elements can vary significantly. Indeed, the value of a pipeline is affected by commercial factors – e.g. sunk costs, price of throughput and availability of alternative routes – as well as non-material benefits and transit fees (Stevens, Transit Troubles: Pipelines as a source of conflict, 2009).

In the energy politics, states act either as risk-averse or risk-prone. Consequently, states are not only concerned about maximizing the profit under various conditions, but also motivated to grip favorable perspective and avoid significant loss. Different risk-taking propensities, therefore, alter the bargaining among pipeline players, affording more risk-prone actors the leverage to drag out negotiations and haggle for favorable terms, disadvantaging the more risk-averse parties and providing the equally risk-averse parties with comparable standing in contracting and payoffs (Omonbude, 2007; Stulberg, 2007)

The weak domestic institutions may challenge the credibility of one state in the eyes of other states. A government with weak domestic institutions can be constrained from below in its capacities not only to formulate and implement coherent foreign policies, but also to pursue effective energy diplomacy. By restricting a government's capacity to marshal specific capabilities and obfuscating stakeholder interests, weak domestic institutions excite the 'bluffer's dilemma', undermining the credibility of that government's policies (Jentleson, 1986).

The logic behind governments' choice to support and forge transit commitments can be explained by two conditions. The first relates to the salience of returns on investment. In the case of transit pipelines, the focus is on receiving 'normal profits' (Stevens, 2009)[14].

This commercial aspect of the pipeline politics is not always the case. As history shows, political geography is tended to change throughout the time, as it

---

14   There is also definition of 'economic rent', where 'rent' is defined as the difference between the full costs of the project (including 'normal profit') and the market price earned by the project. Rent arises because of a monopoly position and/or as the result of a gift of nature where natural resources offer below-average costs of production. Stevens, P. (2009). *Transit Troubles: Pipelines as a source of conflict.* Royal Institute of International Affairs. London: Chatham House.

happened in the Eurasia. Today, owners of the old pipeline system are not the stakeholders, who initiated, constructed and operated those pipelines. In this case returns of investment is not a question of big concern. Instead, there are incentives to bargain for better terms, as the parties risk incurring only opportunity costs and the marginal utility of the operation by disrupting delivery, which confounds the pure economics of post-construction behavior (Stulberg, 2012). Consequently, 'new' transit states, less concerned on recovering financial investments, try to maximize their gains from the transit and arbitrarily change the transit terms. Indeed, 'in cases where pipelines preceded establishment of cooperative political relationships, such as the situation resulting from the Soviet breakup, pipeline and transit relations often serve as a source of friction and contention between states' (Shaffer, 2009).

The second condition is the existence of strong institutions and effective governance in the energy sector. Institutions are crucial in shaping states capacity. Moreover, transparent allocation of property rights (e.g. authority to access, set prices, tax, collect off-take and transport energy resources, etc.) among state and non-state actors is crucial to distinguishing primary stakeholder interests as well as to establishing clear rules for the regulatory game for domestic and international audiences alike (Stulberg, 2012). Hence, regulatory institutions ensure that all partners bear both the costs and benefits, and generate incentives to align policies in order to maximize the value of each partner.

Additionally, unclear distribution of ownership and control among parties compound strategic energy bargaining problems by challenging decision-making on separate issues within a regulatory system. This encourages opportunism by self-interested parties (state and private), raising the transaction costs of organizing and integrating discrete regulatory policies, which aggravate the 'bluffer's dilemma' (Jentleson, 1986).

The empowerment of multiple stakeholders with overlapping authority also damages common understanding of the respective policy interests, accentuating information asymmetries around the preferred costs, tariffs and revenue streams of the respective pipelines for both domestic and international partners and may challenge 'contract stability' causing its renegotiation (Stulberg, 2012).

According to Stulberg, it is possible to make three conclusions based on these analysis:

First, where the main shareholders place a premium on returns on investment *and* bargaining takes place among governments with transparent regulatory systems, the main actors are likely to be well positioned to uphold mutually beneficial and credible commitments to cross-border export. Market and institutional incentives combine to align interests in cooperation, creating conditions conducive for the attenuation of information asymmetries among bargaining partners (Stulberg, 2012).

Second, where the main shareholders are not impelled to recoup investments *and* bargaining takes place among governments with opaque regulatory institutions, the respective pipelines are very likely to be marred by arbitrary disruptions. The incentive to gamble on future commercial or discretionary pay-offs combine with the lack of transparency and lower costs for breaking promises among domestic stakeholders to escalate cross-border pipeline politics (Stulberg, 2012).

Finally, under mixed conditions, where the main stakeholders are either not concerned about the returns on investment or must interact with governments that blur the division of the regulatory authority, the parties must contend with significant but manageable commitment problems. Bargaining problems can be overcome to the extent that cross-border energy transit arrangements can be buttressed by external mechanisms to compensate for either the riskiness or weak/non-transparent regulatory capacity of the parties (Stulberg, 2012).

Pipeline politics should also be reviewed from the credible commitment of the partner parties. Although states' actions may unilaterally cause disruption of energy supply, these actions are, in fact, the product of strategic interaction. Such interaction is multidimensional, involving not only the posturing among the direct parties to pipeline projects, but the shifting sands of markets and domestic institutions that ultimately shape the value of such deals (Stulberg, 2012).

Since all energy players – energy suppliers, pipeline operators, transit states and customers – have to deal with risks and uncertainty, they may possess distinct frames, intuition or benchmarks for assessing the pay-offs of future transit transactions (Mercer, 2010).

3.3.4 Political economy of network interdependence

Access to the adequate sources of energy supply is accepted as the core of energy security. In the case of natural gas, existence of limited supply sources cause dependency of consumer from single supplier by giving leverage to the latter, which may spill over into political, as well as economic arena. Thus, in an interdependent world, energy security and political dependence are inexorably intertwined, particularly when the political preferences, understandings, and objectives of the users and suppliers differ substantially (Ericson, 2009).

Acquisition and transportation of the natural gas requires certain technologies, which have a crucial impact on its economies. Indeed, transportation and storage of the natural gas is difficult and expensive. As a gas, natural gas carries far less energy per unit volume, making road and rail transport, even when it is compressed, uneconomical for industrial use (Ericson, 2009). Its transportation in the liquefied form can be economically advantageous, if transported over long distances. Among all other options, the less expensive way of transportation of natural gas over the long the distance is through a large-diameter pipeline. However, the construction of pipeline network is extremely expensive and in most cases only one pipeline has been built linking the source and the market. This is the reason why gas markets are highly inflexible and political. Even though commercial considerations play an important role and have been used as a pretext while negotiations, it often takes second place to political issues. Inflexible and political nature of the market leads to the formation of interdependence between supplier and consumer.

Furthermore, it creates a true 'natural monopoly', based on long-term commitments of gas purchase, in order to justify the investment and development costs[15]. Consequently, supplier and consumer become linked through a long-term contractual relationship, which is substantially protected from outside competitive pressures. Through the monopoly, supplier gets tremendous market power, where consumer's alternatives are limited.

However, the picture is not one sided. In gas market, both parties are interdependent, as long as both have developed alternatives.

3.3.5 Nature of pipeline dependence

Pipeline plays the role of a unique channel between supplier and consumer along the pipeline. Unavailability of alternatives and tremendously high costs

---

15   But once the pipeline system is in place, with development costs sunk, natural gas can be supplied at a unit cost far below what any competitor would require to set up a competing supply system and it will allow to maintain its monopoly. Ericson, E. (2009). Eurasian Natural Gas Pipelines: The Political Economy of Network Interdependence. *Eurasian Geography and Economics*, 50 (1), 28-57. P-29.

of construction for a new pipeline generates a natural monopoly in the favor of supplier. Moreover, inflexibility of the pipeline structure may also lead to monopsony, where supplier has to deal either with one or few buyers. Such pipeline networks form a "lock-in" relationship, where lack of alternatives imposes non-market bargaining and creates a potential for political leverage to play a decisive role in the high-stakes games.

Another aspect of such relations is 'rent extraction' exercised through price, which is bounded below by unit costs of production plus the low marginal cost of pipeline transport, and from above by the (relatively much higher) costs to the users of retooling to use alternative fuels (Ericson, 2009). Lack of price competitiveness causes a 'political pricing' by reflecting more interstate relations rather than costs and benefits (Stern J. , 2007).

This picture may change when stronger states interact. The price is set on a negotiated formula – netback market value – for natural gas delivery. Reliability turns to be a main factor for supplier. Hence, dependence of supplier from consumer makes the former also vulnerable in terms of supply disruption.

In the case of asymmetrical interdependence, supplier has minor vulnerability, when consumers depend extensively from supplier, because of infrastructure and monopolistic position of the former. This is the case for the Eastern European and CIS countries, which are heavily dependent on Russia in natural gas supply.

The vulnerability of the EU within this context is twofold. The first vulnerability arises in the form of ultimate monopoly of supplier, which is not the big issue. The second and most important is the uncertainty introduced by relations between that supplier and a number of 'transit' states, who are even more dependent than Europe on the Russian and Eurasian supply of natural gas to meet their energy needs.

Indeed, transit countries have commercial interests in maintaining steady supply, as do the upstream and downstream partners. But those interests could be trumped, by more direct perceived threats to their economy, sovereignty, or security (Ericson, 2009).

In the case of gas pipelines, the negative actions will have damaging 'domino effect', disrupting production, paralyzing distribution, challenging investment, consumption, and etc.

3.3.6 Asymmetric interdependence

The asymmetric structure of interdependence, driven from transportation of natural gas through fixed pipelines, occurs due to two reasons. First, disruption may impose different costs on various actors by allowing some to manage these costs more easily than others. The second reason is the level of the control mechanisms and ability to influence the flow and its impact (Ericson, 2009).

If transit states along the pipeline are dependent on the energy resources transported through that pipeline, they become more vulnerable, since they lack levers more than others. Absence of the alternatives may hinder economic development of transit state by having direct impact on production of industrial and commercial users. These factors may also cause political and social instability within the country. That is why, countries dependent from single supplier or pipeline must try to protect themselves, against even short disruptions, through building storages, developing alternative sources of supply and maintaining flexible technologies that allow effective use of alternate energy carriers (Stern J. , 2007).

Although suppliers have direct control of the flow, their control can be limited via alternative sources (in this case LNG) and by-pass pipelines. On the other hand, if supplier is a rentier state by continuous exercise of flow interruption, it will bear financial consequences in the form of income loss and budget deficit. Of course, it does not have a serious impact on economy in the short run. However, in the long-term, if the pipeline disruption and hence lack of substantial income is sustained long enough, exhausting financial reserves, straining the willingness of international markets to continue providing credit, and perhaps triggering serious inflation due to excess monetary emission as a substitute for natural gas revenues (Ericson, 2009).

There are certain intervening variables, which determine the outcome. The market power of supplier should be considered as an important issue. If supplier has a substantial market power and monopolist position, the threat of asymmetrical interdependence will amplify. Consequently, intervention of national governments into the negotiation becomes necessary on the background of such asymmetry in market power. National governments rather than energy firms play an active role in determination of terms of transit and amounts of transported volumes of natural gas. However, state-level negotiations may remain ineffective in softening the asymmetry (Stern J. , 2007).

3.4 Pipeline transportation and supply challenges

The landlocked nature of the Caspian region constrains supply options of oil and gas resources from the region to world markets by leaving less space for maneuverability. In order to export hydrocarbon resources, upstream countries have to get access to transportation facilities in neighboring countries. The compelling fact about natural gas supply by pipelines is that the delivery must involve transit states. In fact, transit lines are extremely vulnerable to political manipulation and economic pressure, which will siphon off any profitability in what is a zero-sum game between the pipeline owner and the transit country (Stevens, 1996). Moreover, due to the high costs, the lengthy time factor in mobilizing finance and building the pipelines, and the geographic limitations on venues, energy-importing and energy-exporting states are limited in their supply venue options, and it takes years to establish alternative routes if a transit state disrupts the supply flow (Idan & Shaffer, 2011).

Transportation of oil and gas from the landlocked remote areas requires a reliable transit corridor that could efficiently serve the requirements not only of upstream countries, but also midstream and downstream countries. Such constellation of the transportation may increase the dependency of landlocked energy exporters on transit states, due to a relative lack of flexibility in finding alternative transportation routes.

Production and export of crude oil from landlocked areas is a quite different process than production and export of natural gas. First, oil can be transported to world markets by pipelines, railway and then by sea tankers. In contrast to oil, transporting natural gas is more expensive and there are only two options for its delivery: pipelines and Liquefied Natural Gas (LNG). Normally, LNG is cost-competitive with pipelines only over distances in excess of 4000 km. That is why pipelines are highly required to deliver natural gas to markets from Azerbaijan and whole Caspian region, while LNG is not cost-effective in a short distance. Second, natural gas delivered to markets via pipelines creates long-term linkages between supplier and consumer. Any interruption to the flow would risk devaluing the entire investment both upstream and downstream of the pipeline (ESMAP, 2003). Compared to natural gas, the case of oil is different. Since there exists a global oil market, the producer can sell its product to any buyer and the consumer can easily shift from one seller to another. So natural gas supply strategy necessitates an accurately measured and market oriented policy.

The natural gas supply from the Caspian region to the European markets will significantly reduce Russia's energy monopoly. However, due to landlocked nature of the Caspian region, the export of natural gas via pipelines from Azerbaijan and Turkmenistan is very complicated. Transportation of natural gas from the Caspian region, namely from Azerbaijan and Turkmenistan, to the Europe will require transit pipelines crossing territories of many states and a pipeline which will be link these two main suppliers through the Caspian Sea. However, transportation of natural gas by pipelines from the region to the European markets is the only relevant export option at the moment.

It is worth to underline that there are political factors challenging the realization of natural gas supply via Trans-Caspian pipeline (TCP) from Turkmenistan to Azerbaijan. Since the TCP has to be constructed through the seabed, it is less likely that Russia and Iran will not object the construction of thay pipeline based on their geopolitical interests, sometimes hidden behind environmental issues and the unresolved legal status of the Caspian Sea. On one hand, Russia prefers to keep control over the hydrocarbon transportation from the Caspian region, and on another hand, Iran will favor supply of Turkmen gas through its own territory. According to Moscow and Ashgabat, the construction of TCP can be possible after final agreement among all littoral states. In this case, considering geopolitical situation in the region, Baku would certainly avoid open confrontation with Moscow and prioritize the development of production in its own reserves rather than support export projects from Turkmenistan. Moreover, apart the political support from some states there are no far no any commercial players who are ready to finance TCP (Pflüger, 2012).

4 Energy and Politics

Since long, the Caspian region due to its geographical location and natural resources was and stays at the center of political and economic interests of regional powers. The region encompasses the territories of five littoral states bordering the Caspian Sea: Azerbaijan, Kazakhstan, Turkmenistan, Russia and Iran. However, when energy politics come into view, only three states of the region, namely Azerbaijan, Kazakhstan and Turkmenistan, are in the focus. The resources of the region are unevenly distributed. Kazakhstan, Azerbaijan and Turkmenistan hold significant amounts of the proven energy reserves in their respective sectors, which put these countries at the focal point of the Caspian energy politics. By contrast, Russia and Iran possess relatively small shares of proven reserves in their sectors of the Caspian Sea.

Chapter 4 describes the kind and dynamics of pipeline politics that started in the mid-1990s in the region, when initial rules of the energy game were set. The Caspian "energy game" is associated primarily with the oil and gas export from Kazakhstan, Azerbaijan and Turkmenistan. Analysis of the first phase of the Caspian energy development (from 1991-2005) will help to track the changes and understand the reasons of the shifts happening in the second stage (from 2006 to present). The chapter first analyzes the energy capacity of the region, and then elaborates on the policies of regional and external actors during the first phase of the Caspian energy development. The role and interests of Iran and Russia are elaborated within the context of geopolitical rivalry, as both energy producers have relatively small shares of oil and gas supply from the Caspian and their concerns in the region's energy politics is strongly linked to their global, regional and foreign policy directions and priorities.

4.1 Caspian energy development in 1990s

Development of the oil and gas industry in the Caspian region followed through different trajectories determined by political and geographic conditions, which

© Springer Fachmedien Wiesbaden GmbH, part of Springer Nature 2018
S. Amirova-Mammadova, *Pipeline Politics and Natural Gas Supply from Azerbaijan to Europe*, Energiepolitik und Klimaschutz. Energy Policy and Climate Protection, https://doi.org/10.1007/978-3-658-21006-9_4

divides Caspian energy development into two phases. Priorities given to the oil and gas supply have been shaping the nature of the energy politics in different periods of time. Various political factors and economic conditions have determined the trajectory of each phase and influenced the kind of the energy politics in the region by shifting priorities and the level of interdependencies among the stakeholders. Phase one illustrates the energy politics between 1991 and 2005, when development of the oil production was set as a priority (Shaffer, 2010). In the first decade after collapse of the Soviet Union, international oil companies and governments engaged in the development of hydrocarbon resources with primary focus on oil production, partially due to the greater capital expenditures necessary to start up natural gas production (Bahgat, 2007). Phase two covers the period from 2005, when the economic situation in Azerbaijan and Kazakhstan began to change and oil production start bringing revenues to producers (Shaffer, 2010). This phase is strongly linked to the rise of natural gas production and its export to the energy markets.

A number of factors played an essential role in shaping the first phase of Caspian energy developments, especially geography. It was playing a crucial role in development of Caspian energy production and export. Azerbaijan, Kazakhstan and Turkmenistan are the landlocked countries with limited export options depending on neighboring transit states for transportation of goods[16]. In other words, landlocked countries need to attain and maintain access to infrastructure and facilities in neighboring states for transit, in order to be able to participate in international trade (Idan & Shaffer, 2011, p. 241). The landlocked nature of the region was constraining transportation options of fossil fuels, as landlocked energy producers tend to have quite different patterns of production and export than those that border the sea, in areas such as export routes, price and international involvement (Shaffer, 2010). A need for construction of new pipelines, the fact that export will have to be transited through the territory of other states and the costs[17] of such shipment were not commercially attractive for the companies. In case of realization, pipelines can become the subject of political manipulations and making the transportation vulnerable to disruptions.

---

16   Defining terms: The term landlocked refers to a country having no seacoast whatever, being completely Mediterranean. The term transit state is defined as any state with or without seacoast, situated between a landlocked state and the sea, through whose territory traffic in transit passes. For more details see Glassner, M. (1970). *Access to the Sea for Developing Land-Locked States.* The Hague: Martinus Nijhoff.

17   Realization of the energy shipment from the landlocked regions is significantly more expensive than doing so from states with sea access.

Furthermore, landlocked nature of the region puts additional constrains on the states narrowing their foreign policy options and leaving a little space for political maneuvers. At the same time, it affects energy partnership with consumer states and forces landlocked ones to pursue more balanced foreign policy with neighboring and partner states.

Caspian energy producers for a long time have been dependent on the Russian transportation network, whereby oil and gas resources of the region were exported to the world markets through the old soviet pipeline system in early years of independence. Dependency on Russian transportation network was restricting the sovereignty of post-Soviet states by maintaining Moscow's direct influence on economic and strategic decisions. As the history proves, the land-locked post-Soviet states, namely Kazakhstan and Turkmenistan, were much vulnerable to Moscow's military sanctions and changed their political paths when Russia applied economic sanctions[18] (Idan & Shaffer, 2011, p. 246). Therefore, dependency of Caspian producers from transit states, primarily on Russia on one hand, and vulnerability of energy shipments to the disruptions and political manipulations, in general, on the other hand, was raising questions of security for the potential investors. The landlocked nature of the Caspian region and the fact that export of the resources have to be transited through the territory of other states, in the light of growing political instability and security concerns, were introducing additional risks and making energy projects from the region less attractive for the foreign investments. In addition, energy transportation from the landlocked region requires more financial resources for realization. In order to start and realize such projects political support was necessary.

Following the dissolution of the Soviets, the USA was very active in the Caspian region and highly committed to fostering the independence and security of the new post-Soviet states. Washington viewed the creation of new energy export projects as the means to establish the new states' independent security orientation and links with the United States and Europe (Shaffer, 2010). Ankara and Washington were promoting development of the new oil and gas pipeline

---

18   Turkmenistan and Kazakhstan were and still are more dependent on transit through Russian territory, which enables Moscow to keep control over the energy export from these countries. Despite Ashgabat several times expressed its interest in trans-Caspian export pipeline and very ready to take concrete steps, it never happened, because of disruptions and cut-offs caused by Moscow regarding the export of Turkmen gas via Russian territory. Kazakhstan also has refrained from building a trans-Caspian oil export pipeline that would join BTC line, bowing to Russian pressure.

routes from the Caspian in the western direction. Washington became involved in promoting new east–west routes from the Caspian in order to lower the perceived risk involved in these complex projects and help them materialize along the east–west routes that the U.S. favored (Shaffer, 2010). In fact, proposed export projects were very costly and was driven from political rational aimed to minimize Russia's influence in the region. Washington's involvement in the realization of the energy projects provided political support for the international oil companies by guaranteeing success of the projects. Involvement of the US and political support from the Western allies helped to lower the risks linked with the realization of the complex projects and get the investment to materialize projects along the southern energy corridor (Pflüger, 2012).

At the beginning several oil and gas pipeline projects with participation of Azerbaijan, Kazakhstan and Turkmenistan were proposed. The main objective was to establish multiple pipeline routes, in order to lessen dependency of the regional states and European consumers on Russia. In the Soviet period, the region's infrastructure connected the area's oil and natural gas production mostly to Russian territory (Shaffer, 2010). The huge support was given to the construction of the oil and gas pipeline through the Caspian Sea. Conversely, due to Moscow's plans and vision on the means of exporting region's hydrocarbon resources, the outcome was mixed and they could not reach all initial objectives. From the beginning, Moscow was blocking all pipeline projects crossing through the seabed by referring to environmental concerns and unresolved status of the Caspian Sea. Moreover, whilst Washington viewed the region as possessing important strategic importance and wanted to set conditions that would consolidate its connection to the new states in the region, Russia and Iran sought to serve as transit states in order to have a lever to influence the strategic orientation of the region (Shaffer, 2010). Hence, the initial attempts to construct Trans-Caspian Pipeline system for exporting Kazak oil and Turkmen gas bypassing Russia and Iran failed at the end of 1990s.

Geopolitical rivalry taking place among regional and non-regional actors ended with the development of two storylines in Central Asia and Caucasus, associated with the establishment of Caspian-Pipeline-Consortium (CPC) and Baku-Tbilisi-Ceyhan (BTC) respectively. CPC was built from Kazakhstan through the Russian territory, with the strong encouragement of the U.S. government and major investment by American oil companies. It was not possible to fully reduce dependency of Central Asian energy producers on Russian transportation system, whereas with the BTC line Azerbaijan achieved independence in energy export (Chufrin, 2001; Bilgin, 2009).

Azerbaijan and Kazakhstan were pursuing multiple pipeline policy during the first phase. Both were exporting oil to Russia and other neighboring countries through different pipeline systems. The decision on how and where to export were influenced mainly by political considerations. Being located in geopolitically complicated region, the interests of Russia, Iran, Turkey and the United States were considered. The decisions were affected more by strategic rather than economic considerations.

Along with commercial interests and expectations, geography, political objectives of all parties and unresolved status of the Caspian Sea were determining the nature of the energy politics pursued in the Caspian region. In fact, development of the oil production has had significant impact on commercial and political dynamics of the region.

4.2 Assessing energy capacity of the region

The Caspian region is one of the world's oldest energy provinces (Campbell, 1997) and the history of the oil production in the region dates back to the nineteenth century. Rich hydrocarbon resources have always made the Caspian Basin attractive for regional and non-regional actors. The first oil drillings in the region were commenced in Azerbaijan in mid-nineteenth century (Campbell, 1997; Mairet, 2006). By 1900, Baku was producing about half of the world's total crude oil (Bahgat, 2002). This impressive level of production was, to a large extent, the result of combined efforts and investment by the Noble brothers, the Rothschild and Royal Dutch Shell, who helped Tsar Russia to develop Caspian oil resources (Forsythe, 1996). Starting from the early twentieth century, control over oil resources had strategic importance for the parties competing for the power during the World War I and II (Bahgat, 2007).

Caspian oil production was delayed and substantially reduced during 1950s. Due to the lack of technology to explore offshore areas in the Caspian Sea, exploitation of the onshore and near shore fields became prioritized by the Soviet Union. Once Azerbaijan's onshore and near shore oil was developed, Kremlin mobilized its huge resources for the development of other onshore areas such as the Volga-Urals region and West Siberia (Effimov, 2000; Bahgat, 2003; Goldman, 2008). Consequently, investment in the exploration of the new fields in the territory of Azerbaijan was postponed with the discovery of the oil fields in Volga-Urals and West Siberia by Soviets (Effimov, 2000; Bahgat, 2002).

The situation was different in Kazakhstan. The scarcity of modern drilling technology to control early blowout[19] was challenging the exploitation of the Tengis oil field (Kleveman, 2003). On the other hand, Kashagan was protected as natural reserve and therefore Kazakhstan's huge energy potential was poorly developed during the Soviets. It was clear that most of oil reserves are situated in the offshore areas. Shifted investment priorities in Soviet Union and lack of the required technologies for the development of deep-water offshore fields resulted with the low development of the Caspian energy capacity. Moreover, this policy led to decreased exploration and production of oil in the Caspian for most of the second half of the 20th century (Bahgat G. , 2007).

In contrast to oil, in the case of natural gas the financial, technical and engineering concentration was put on the development of the energy resources in Turkmenistan and Uzbekistan (Shaffer, 2010). Due to the lack of technology natural gas reserves of these countries were not developed fully. Therefore, energy capacity of Azerbaijan and Central Asian countries were not accurately estimated and explored during the Soviet era. Hence, existed ambiguity and lack of appropriate assessment end up with the rise of various myths around the Caspian energy bonanza in the early years of post-Soviet era. It was largely believed that the Caspian Sea contains large resources of oil and gas capable of much greater production than actual (Gelb, 2005).

4.2.1   Myth and reality of the Caspian energy

During the 1990s, estimates and projections dominated the debates about the energy capacity of the three Caspian states. Lack of accurate assessment of the region's recourses led to numerous myths about the energy potential of the regional states. Some of the top officials in Kazakhstan and Azerbaijan described their countries as "another Middle East", "another Saudi Arabia" and

---

19   A blowout is an uncontrollable increase of pressure, oil comes rushing upwards as may be seen an old black and white films - a spark and oxygen are enough to enflame the gases and make the well spit fire. Kleveman, L. (2003). *The New Great Game: blood and oil in Central Asia.* New York and London: Atlantic Books. Mairet, F. (2006). *New Stakes in the Caucasus and Central Asia: Caspian Energy Resorucs and International Affairs.* Bloomington: AuthorHouse..

"another Kuwait". Even data provided by energy firms like British Petrolium, or US Department of Energy, or Oil and Gas Journal at the beginning of the 2000s had huge gap, and great disparity between each other on the potential oil and gas reserves of the three Caspian states (See Table 1) (Bahgat, 2002).

Table 1:   The Caspian Proven Oil and Gas Reserves[20]

Country	BP		Oil & Gas Journal		US Department of Energy	
	Oil	*Gas*	*Oil*	*Gas*	*Oil*	*Gas*
Azerbaijan	6.9	30.0	1.2	4.4	8.0	11
Kazakhstan	8.0	65.0	5.4	65.0	13.2	68
Turkmenistan	0.5	101.0	0.5	101.0	1.7	126.5
Total	**15.04**	**196.0**	**7.1**	**170.4**	**23.5**	**205.5**

Source: BP 2001, Oil and Gas Journal 2000 and Energy Information Administration 2000

There were huge disparities among prognoses provided by state agencies and oil companies. Some were comparing the region with the Middle East or Saudi Arabia or Kuwait (Bahgat, 2002). Moreover, some analysts argued that the data was exaggerated by Washington in order to reduce American and Western energy dependence on the Persian Gulf and to justify political support for international oil companies in the post-Soviet countries (Nanay, 1998). Hence, politics rather than economics were playing a main role in the development of the energy production and export projects during early years of political independence of the Caspian states.

In fact, Azerbaijan, Kazakhstan and Turkmenistan possess huge oil and natural gas fields though with different capacities, which shape energy politics of these countries and determine their role in the energy game. Kazakhstan has the region's largest crude oil reserves and its production accounts for approximately one-thirds of the region's overall oil output . The main oil fields – Tengiz, Kashagan and Kurmangazy – have been developed by international oil companies starting from 1992. Although the Tengiz field was originally discovered in 1974, the full exploitation of the field started after 1993 when Chervon and Kazakhstan signed an agreement on a joint venture. Kashagan is the largest oil field outside the Middle East and the fifth largest in the world (EIA, 2015). Kazakhstan's third largest oil field Kurmangazy is developed based on a

---

20   proven oil reserves in billion barrels and proven gas reserves in trillion cubic feet (Tcf)

production-sharing agreement between Russia and Kazakhstan, since this field is situated on a maritime border of these two countries.

The region's second largest oil producer is Azerbaijan. The oil reserves of the country constitute about 20% of current regional crude oil output (Gelb, 2005, p. 1). The main oil production comes from Azeri-Chirag-Gunashli mega-structure, which makes country's more than 80% of total oil output (EIA, 2015). The field contains an estimate of 5.4 billion barrels of recoverable crude oil (Mairet, 2006). However, Azerbaijan's future oil prosperity is highly uncertain, due to several disappointing drilling results. For example, in the mid-2004 and 2005, ExxonMobil and Lukoil failed to discover commercially viable oil reserves at the Zafar-Mashal and Yalama blocks (Bahgat, 2007). Besides, Azerbaijan's oil production started declining from 2010. Whereas oil production in 2010 was 1.0 bbl/d, it was falling to 932,000 bbl/d in 2012, and 910,000 bbl/d in 2013 (EIA, 2015).

Unlike Kazakhstan and Azerbaijan, Turkmenistan does not possess huge oil reserves. Most of Turkmenistan's oil reserves are located offshore and Garashyzlyk onshore area in the western part of the country. Since the political independence two issues have been restraining oil production in the country. First, following the break-up of the Soviet Union, Ashgabat failed in implementing quick economic and political reforms, questioning the guarantee of stable business environment for foreign investments (Bahgat, 2003). Second, the unresolved status of the Caspian Sea and territorial dispute between Turkmenistan and Azerbaijan prevented the exploration of the Serdar/Kyapaz oil field. Ashgabat, however, succeeded in attracting some investment for the development of the offshore Cheleken project. Proved and probable reserves in the contract area are around 3 million bbl of oil and 3 Tcf of natural gas.

Oil production rather than gas was the main focus of the international companies who entered the region in 1990s, although, natural gas reserves of the region have a higher share in the world's total natural gas than oil. According to BP statistics, the region's share of proved natural gas reserves in the world is 11.3 percent. The biggest discovered natural gas fields are onshore in Turkmenistan, Kazakhstan, and Uzbekistan, as well as offshore Azerbaijan. Russia and Iran also have sizable natural gas deposits on their offshore areas of the Caspian Sea. According to EIA around a quarter of Kazakhstan's gas production comes from onshore fields and the eastern part of the country and majority of the Turkmenistan's gas production comes from onshore areas and fields in the southeast of the country.

Turkmenistan is by large the predominant producer of natural gas in the Caspian, with about 75% of the region's total gas output (Gelb, 2005). According to independent audits conducted in 2008 and 2009, Turkmenistan largest natural gas field is the South Yolotan-Osman (renamed to Galkynysh in 2011), which is the world's fourth largest natural gas field (EIA, 2015).

Natural gas reserves of Kazakhstan constitute associated gas found in the oil fields. The Karachaganak field is considered one of the world's largest gas-condensate fields, with 46% of natural gas. Other oil fields of the country have about 10% of gas condensate (EIA, 2015). Despite the large natural gas reserves, Kazakhstan imports natural gas from neighboring countries to meet its gas demand. The reason of turning Kazakhstan to a gas importer is twofold. First, about 70% of its dry natural gas production has been re-injected into oil fields to enhance oil production (EIA, 2015). The second is determined by the lack of domestic pipeline infrastructure, linking the production area in the west with the demand centers in the northern, southern and eastern parts of the country.

Unlike other Central Asian producers, natural gas production in Azerbaijan comes only from offshore areas. In 1999 Azerbaijan experienced a positive development with the discovery of the Shah Deniz field. Most of the natural gas production comes from that field. Some remarkable volumes of natural gas are produced alongside oil in the ACG field. With the development of the Shah Deniz field Azerbaijan has become another main natural gas producer of the region, and a new stage of the energy politics has started making natural gas supply from the Caspian to the world markets more realistic. This development and the proportion of the natural gas reserves in the region imply that the future of energy politics in the Caspian primarily will be linked with the future growth of natural gas production.

Table 2:   The Caspian Proven Oil and Gas Reserves[21]

Country	BP		Oil & Gas Journal		US department of Energy	
	Oil	*Gas*	*Oil*	*Gas*	*Oil*	*Gas*
Azerbaijan	7.0	31.5	7.0	35.0	8.5	50.0
Kazakhstan	30.0	45.7	30.0	85.0	30.0	85.0
Turkmenistan	0.6	618.1	0.6	265.0	1.9	265.0
Total	**37.6**	**695.3**	**37.6**	**385.0**	**40.4**	**400.0**

Source: BP 2012, Oil and Gas Journal 2012 and Energy Information Administration 2013

---

21   proven oil reserves in billion barrels and proven gas reserves in trillion cubic feet (Tcf)

When the exploration works started in the seawaters of Azerbaijan and Kazakhstan, there were disappointing results from drillings, but also unexpected positive discoveries in Azerbaijan and Kazakhstan. In 2001 ExxonMobile suspended all operations in the Oguz field of Azerbaijan because the well it had drilled turned out to be dry. Also Chevron stopped working on the Absheron field off Baku because only a very thin layer of gas was discovered (Bahgat, 2002). But parallel to these, there were unexpected positive discoveries in Azerbaijan and Kazakhstan, particularly Shah Deniz and Kashagan. All these events led to a reassessment of the energy potential of the Caspian Basin during the late 2000s, which also changed the nature of energy politics in the region (See Table 2).

The new figures prove that the energy potential of Azerbaijan, Kazakhstan and Turkmenistan are less than those of the Middle East. That means the Caspian region does not replace Persian Gulf as a main reservoir of world oil. However, the Caspian region plays an important role in world energy politics, and has a potential to enhance global energy security and diversify existing energy sources in the global market. Current estimates show that proven reserves of the Caspian Sea can be compared with the North Sea and composing about 2.5 percent of the world's proven oil reserves (BP, 2014). Moreover, oil supply from the region to the global market in the year 2013 was around 3.4 percent of world's total crude oil supply (BP, 2014; IEA, 2015).

4.3 Geopolitical dimension: energy interests and foreign policies of external actors

Following the end of the long-lasting Cold War and the breakup of the Soviet Union in 1991 the geopolitical situation in the post-Soviet area has dramatically changed and the Caspian Sea region has risen from relative obscurity to considerable prominence in international affairs (Chufrin, 2001). For decades, Iran and Russia dominated the southern and northern parts of the Caspian Sea and were the only strategic powers in the region (Ehteshami, 2004). A "power vacuum" emerged after the collapse of the Soviet Union triggered jockeying for influence by regional and non-regional super/major powers and transformed the Caspian region into an area of huge competition over the control of energy resources. This competition over the security and control also aimed to counterbalance existing traditional Iranian and Russian power in the region. A new geopolitical situation formed during the 1990s characterized with the increased involvement of state and non-state actors, and at the same time, political competition between Russia and the West. Security and energy became key issues determining the nature of the strategic setting of the Caspian region.

Next to the decline of Russian influence in the region, another "window of opportunity" opened for the international actors in 1990s. The Soviet breakup and subsequent establishment of the new states in the Caspian Basin created an opportunity and sparked international commercial and political interest in developing the oil and natural gas riches of the region. Azerbaijan, Kazakhstan and Turkmenistan had weak economy after the collapse of the Soviet Union and needed foreign investment to utilize their hydrocarbon recourses. Many international oil companies became involved in the operation works in the Caspian region.

Actors interested in the exploration of the oil and gas fields in the region along with new independent states opened a new page of the energy politics in the Caspian, and another phase of the regional rivalry began. Russia, Iran, Turkey, Japan, China, USA and the EU became engaged in political and social-economical dynamics of the region at different levels. The Caspian region has become internationalized to an extent not seen before, and major reconfiguration of power and influence have taken place (Jonson L. , 2001). Russia dominated energy transportation, since shipping of oil and gas to markets has been implemented through the old network. Non-regional actors has very limited power to influence politics and economics in the region. Therefore, an intense struggle started over the control of new pipeline routes, which will connect the oil and gas fields in the Caspian Basin with the energy markets. Western states put all efforts to diminish dependence of the Caspian states from the Russian infrastructure for transportation of hydrocarbon resources (McCarthy, 2000). Particularly the United States and the Turkish government pursued a clearly aligned policy to support development of the oil and gas industry in the newly independent states of Central Asia and the Caucasus. The main objective was the construction of new east-to-west oil and gas pipelines thus would enable oil and gas producing countries to export their resources without traversing the territory of the Russian Federation or Iran (Pflüger, 2012). The idea was supported by the US State Department, and also strong emphasis was put on common cultural, historical and linguistic links of the regional countries with Turkey, which should assure the success of the pipeline projects.

Strategic and economic interests of Russia, Iran, US and Turkey led to the formation of a political game sketches, where the parties were playing together and, at the same time, against each other. The first stage of the Caspian energy development was mostly intertwined with political considerations of producers, as well as regional and external actors. The energy and pipeline politics started in 1990s were described as a new "Great Game" of the 21st century. In fact, external actors have different levels of influence in shaping the nature of the energy politics in the Caspian region. Some play a more active role, some are passive and other has only limited capacity to influençe the direction of the pipeline politics. An analysis is important to understand their actions and motives in the second stage of development.

4.3.1   United States

The location of the Caspian, between Russia and Iran, determined the U.S. focus on this region (Nanay, 1998). The situation formed after the collapse of the USSR followed with the changes in the geopolitical map of the Eurasia, created an opportunity for U.S. to enhance its political, economic and energy interests in the Caspian region. Opening of the region's energy reserves for foreign investments, which were closed for decades during the Soviet era, economic weakness of the littoral states and interest of American oil companies in the hydrocarbon reserves of the region brought the U.S. to start more active foreign policy.

Geopolitical and energy security concerns determined Washington's interests in the region. Throughout of the first phased of the Caspian energy development three directions constituted the key points of US' regional policy in the region. In order to strengthen economic and political independence of the region's new states from Russia, it supported state-building process in each country. Hence, mainly in the South Caucasus U.S. was trying to facilitate conflict settlement processes. Second U.S. was interested in developing region's energy resources and construction of the new pipelines from the Caspian to the world markets. This should decrease dependence of the Western states on oil imported form the Persian Gulf, on one hand and diminish Russian monopoly over transportation networks from the region, on the other hand (Bahgat, 2003; Haas, 2006). New pipelines had to minimize supply risks and reduce Russia's economic and strategic influence in the region, which was its key competitor in the Caspian region. Third, the involvement in the Caspian was a part of Washington's foreign policy strategy. In the light of U.S. sanctions against Iran, through active involvement in the Caspian official Washington sought to isolate and limit Tehran's role in developing new energy transportation infrastructure in the Caspian region. Despite the advantages of the transportation of the oil resources through Iran, the pipeline proposals were not approved by the state administration[22]. Support for the construction of the new pipelines, which bypass territories of Russia and Iran had to limit political influence of the traditional regional players in the newly independent states. The US has repeatedly maintained that its policy is aimed at breaking the Russian monopoly over energy transportation routes, but that it is not anti-Russian in itself (Haas, 2006).

However, the outcomes were different and in reality, various trajectories were drawn. Azerbaijan and Kazakhstan were the key countries where U.S. followed distinguish foreign policies. Turkey's involvement in developing the energy transportation infrastructure from the Caspian region to the global markets was vigorously supported by U.S. that lead to formation of strategic and economic partnership particularly with Azerbaijan. Sitting in a key location in the Southern Caucasus and bordering on Iran, Azerbaijan became the pivotal country for the US government's investment advocacy agenda (Nanay, 1998). A large number of the offshore and onshore contracts were signed with Azerbaijan during the 1997-1999. Apart developing oil and gas fields, Washington was interested in setting stable political environment in Azerbaijan, which was necessary to create a regional pipeline hub with oil and gas export routes running from Caspian Basin through Georgia and Turkey.

With realization of BTC oil and SCP gas pipeline projects the objectives were met. Within the context of an east-west energy corridor, long-term goal with these projects were to create an energy source that are independent of Russia and Iran and emanated from countries that considered US allies – Azerbaijan, Georgia and Turkey. Oil transported from Caspian via BTC line has been viewed as enhancing the diversity of non-OPEC supply sources, which on the other hand composes U.S. global energy security concerns.

---

22   U.S. oil companies believed that the southern route via Iran could provide the easiest, fastest, and cheapest route to transport Caspian oil and gas. However, U.S. government was against it, which indicated that there was a contradiction between the U.S. commercial interests and strategic ones. Washington's rationale is that aside from sanctions, there could be security of supply concerns related to oil transported via Iran. Caspian oil sent via the Persian Gulf would be subject to the same potential bottleneck as much of Middle Eastern crude if flows were disrupted through the Strait of Hormuz. Bahgat, G. (2003). *American Oil Diplomacy in the Persian Gulf and the Caspian Sea*. Gainesville, Fl., USA: University Press of Florida.

Involvement in Kazakhstan has been determined mostly by energy and commercial factors. American oil company Chevron signed the region's first onshore joint venture for what is even today considered to be one of the world's giant oil fields – Tengiz. Conversely, it was not possible to construct independent pipeline from Kazakhstan through the Caspian Sea within the framework of east-west energy corridor. In Kazakhstan Russia was considered as partner rather than competitor in constructing CPC pipeline, the flagship pipeline project in the Caspian region. However, geopolitical competition with Russia and issues have posed by tanker transit through the Bosporus have led to the serious study of other export options (Nanay, 1998).

Parallel with geopolitical importance of the Washington recognized geo-economic importance of the Caspian earlier than Europe (Haas, 2006). By promoting new transportation routes and following multiple pipeline politics U.S. intended to diversify supply sources to Europe and decrease its NATO allies dependency on Russia. This should limit leverages used by Moscow to affect political decisions in Europe.

4.3.2 Russia

The Caspian region represents a significant region for Russia in geopolitical, strategic and commercial terms. In fact, energy and pipeline politics pursued by Russia in the region, during the first stage of the Caspian energy development should be reviewed within the context of Russia's national energy security framework with a direct impact on its national economy. While analyzing regional dynamics and strategic competition in the South Caucasus and Central Asia the realist approach becomes especially relevant. Energy politics pursued by major powers reflect their global energy security priorities. In the case of Russia, energy policy has been built on resource nationalism. For years Moscow was controlling not only energy resources of the Eurasia, but also had a control over the pipeline routes, which connected resources with the international markets. For Russian political elite control of energy resources and pipeline network is fulcrum for its power restoration and the key to realizing competitive advantages abroad (Cohen, 2009; Mankoff, 2009). It is also important to note that the oil and gas industry occupies a prominent position in the Russian economy as it generates a substantial share of its foreign revenues.

The breakup of the Soviet Union created a number of different, new stake-holders in pieces of that supply system, and indeed a number of formally independent producing states. Almost all export pipelines still ran through Russia and were controlled by Gazprom, at least to Russia's western border. In particular, almost all Central Asian and Caucasian natural gas and oil had to initially pass through the Russian pipeline network to reach foreign customers (Ericson, 2009). Nonetheless, Russia is not only player in the Caspian region conducting pipeline politics. Other actors involved in their own designed pipeline politics, which has negative effect on Russia's energy ambitions in the region and new pipelines bypassing Russia are posing direct threat to its power in the region (Trenin, 2009). Active involvement of U.S. and other Western countries in the region has been negatively viewed and considered as an attempt to marginalize Russian influence and power in its near abroad. Strategic gains, rather than economic profits constitute the logic behind Russia's pipeline politics. Indeed, economic rational of these strategies is weak and very limited (Vatansever, 2010).

The primary objectives of Russia's regional policy is regain and strengthen position in the whole post-Soviet space, not only in the South Caucasus and Central Asia. Through using diplomatic, economic and military leverages and maintaining control over the pipeline infrastructures Moscow has sought to limit involvement of the external actors in the region. Furthermore, to reassert its influence in the regional states, Russia was playing the role of mediator and even arbitrator in solving disagreements between the conflicting parties, which gave the additional political leverage in former Soviet states. Besides, Moscow was able to influence energy politics of Kazakhstan, Azerbaijan and Turkmenistan during the first phase of the Caspian energy development in a following ways: as an investor or partner in field development and pipeline projects; as a transit country for their exports to formal Soviet republics and other markets; as a competitor in most of these markets; and as a market in its own right (IEA, 1998; Bahgat, 2002).

Thus the Caspian states remained interlinked with Russian in political and economic terms, and Moscow was able to influence and shift energy politics in the region in its favor. In 2000, when Vladimir Putin replaced Boris Yeltsin as a president of Russian Federation, Moscow's Caspian policy became more coherent and aggressive. In order to optimize its strategic and commercial interests in the Caspian, Putin was urging Russian oil companies – Lukoil, Yukos, and Gazprom – not only in becoming involved in the Caspian energy development but also to take a more cooperative stance toward investment from other countries to help promote a higher level of total development (Bahgat, 2003). Some analysts believe that Russia's Caspian policy under Putin moved away from trying to contain U.S. expansion in the region in favor of "constructive engagement" with American government and oil companies (Baker, 2000). By standing out against this, some described the pipeline politics in the region as a "commercial realpolitik", where interests of governments and firms encounter, when the former review the dynamics through political rational and latter from business (Abdelal, 2010).

Summarizing Moscow's policy in the Caspian region during the first phase, it is possible to distinguish two different directions. Kremlin's energy politics was build on competition and partnership. First, in comparing with United States, Russia had more political, economic and military levers to influence and shape the nature of energy politics in the region. In this way, it could not only change the situation in its own favor or even impede development of the new pipeline projects, as it was case in planning of the Trans-Caspian pipeline system. Next, Moscow was increasing bilateral energy partnership with Turkey, another active player in region, and also became involved in the energy development and transportation projects along with the western oil companies by engaging national oil and gas companies. As it can be seen, starting from the late 1990s Russia has succeeded in negotiating and constructing a number of pipeline schemes between its ports and oil and gas fields in the Caspian and its oil and gas companies were heavily involved in most multilateral energy consortia in the region (Bahgat, 2003).

4.3.3 Turkey

Ankara's interests and foreign policy conducted in the Caspian region have been driven by cultural, linguistic and historical ties with the regional states. Following the collapse of the Soviet Union, Turkey sought the ways of increasing its influence in the South Caucasus and Central Asia as part of its regional strategy policy and commercial interests. Given that Ankara was the first trying to build up political, economic and energy partnership with the Caspian states[23],

---

23   Starting from early 1990s, Turkey launched a campaign to establish good contacts with former Soviet republics possessing a great ethnic Turkic population like Azerbaijan, Kyrgyzstan, Turkmenistan, Uzbekistan and Kazakhstan.

in this way constituting itself as a main regional rivalry to Russian and Iranian objectives in the region. Supporting construction of the new pipeline projects through its territory, Turkey pursued not only to become the bridge between energy producing countries of the region and the world market, but also increase its role as major regional player between Europe and the Caspian. Ankara's active involvement in the development of the region's hydrocarbon reserves and transportation projects has been stipulated by commercial as well as political rationales.

Besides, Ankara was the key regional partner of the West in the region. By strengthening political, cultural and economic ties with Azerbaijan, Ankara became involved in negotiation process between Baku and Western oil companies during the pro-Turkish Elchibay's presidency. Despite the *coup d'etat* followed with the renegotiation of the oil agreements between Azerbaijan and oil companies and inclusion of Russian companies to the deal, Turkish companies were able to get shares in Azerbaijan International Operating Company (AIOC). Without a doubt, Moscow and Ankara were considering each other as a direct threat to their energy and regional policy in the Caspian. The struggle for power, especially over the export routes were going through differ passes in the Central Asia and South Caucasus.

Turkey, with the political and financial backing from United States, was more successful in conducting regional policy in the South Caucasus rather than in the Central Asia. Similarly, it was supporting development of the Georgian economy and building good relations with Tbilisi to strengthen its position in the region. However relations with Yerevan were and stay tense because of genocide issue of 1915 and Nagorno-Karabakh conflict as Turkey continued supporting Azerbaijan and keeping closed borders with Armenia. Hence, Azerbaijan and Georgia were the regional partners of Turkey, which end up with of the realization of two major projects, the BTC oil pipeline and the BTE gas pipeline. The development and consolidation of the energy cooperation between Azerbaijan, Georgia and Turkey during the first phase of the Caspian energy development was the beginning of strategic trilateral partnership. That partnership has led to the development of strategic relationship closely aligned in terms of foreign, economic and security policies of these three states. Certainly, the role of United States as a key supporter of these partnership was crucial for the successful collaboration.

Despite successful partnership on the western side of the Caspian, Turkey was not such successful in building partnership with Kazakhstan and getting

Tengiz oil transported through Turkey supported new pipeline system. In August 1995, Turkish Prime Minister Tansu Ciller engaged in a last ditch effort to sway Almaty to opt for a Turkish pipeline. The talks failed as Kazakh president Nursultan Nazarbayev insisted that Kazakh oil should be transported to Russia's Black Sea terminal, which lead to the formation of the CPC in 1996 (Akiner, 2004).

The tied partnership with Caucasian states and realization of the east-west transportation route shaped the current geopolitical map of the region by increasing Turkey's geopolitical position not only in the Caspian Sea region, but also in Black Sea region. Moreover, pipeline politics pursued in the region had to transform Turkey into the ultimate pipeline collector for both oil and gas lines, where it would become an important transit corridor for transferring not only hydrocarbons from Caspian, but also from the Middle East to Europe. On the other hand, it would increase state revenues. By turning itself into the predominant energy transit country, Turkey pursued to become vital energy player at the international level and at the same time to ensure political and economic stability within the country.

4.3.4 Iran

The breakup of the Soviet Union and changes at the political map of the region created both opportunities and new threats for Iran. Just like other regional players, Iran sought to gain influence in the Caucasus and Central Asian states. As soon as regional states regained their independence Tehran mobilized all its efforts to strengthen political and economic relations with them. Iran saw itself as the nexus between the Caspian Sea and the Persian Gulf, which could obviously link the untouched markets in Central Asia and the Caucasus with outside world (Namazi & Farzin, 2004). It is geographical location between the Caspian and the Persian Gulf and relatively developed internal transport infrastructure offered the potential to play an important role as a bridge for the landlocked Caspian states (Bahgat, 2002; 2003). Offering shortest and cheapest route to global markets for oil and gas from the Caspian to the world markets, Tehran was proposing realization of the new transportation projects through its territory. In reality Iran's commercial interests was going beyond just offering its territory for transit of Caspian oil and gas. Undoubtedly possible new pipeline projects and potential income that might be derived from the transportation projects were an important consideration. However, Iran's former president Rafsanjani was looking in transferring the country into the regional hub, which will turn Iran into a gateway to Central Asia and Caucasus (Morady, 2011).

Pursuing active Caspian strategy Iran became engaged in the geopolitical rivalry involving regional and external actors, which composed additional threats to Tehran regional politics. By attending the new Great Game Tehran was particularly concerned about two possible threats — being excluded from a future Caspian Sea grouping (similar to the Gulf Cooperation Council in the Persian Gulf) and the danger of hostile foreign penetration into the region determined in terms of U.S. Caspian policy to sideline Iran in the development and transportation hydrocarbon resources of the Caspian Basin (Bahgat, 2003; Namazi & Farzin, 2004). Based on similarities between Caspian region and Persian Gulf Iran formed its regional policy in a way to contain Western and Turkish influences. Like Moscow, Tehran was not also happy with the enlargement of the relationship between regional states and external actors. Active involvement of U.S. and Turkey in the region's economic and political life was seen as a threat to the state's national and security interests. Being concerned with the military asymmetry emerged as a result of militarization of the Caspian Sea[24], heavy investment in the regional states and especially growing American military role (directly or through a third party such as Turkey), Moscow and Tehran have forged a strategic alliance to resist what they perceive as "American hegemony" in the Caspian and worldwide (Bahgat, 2003).

On the other hand, Iranian politicians see the formation of the independent Republic of Azerbaijan as a threat, which could lead to secessionism and extremist ethno-nationalism with the negative consequences for the nation-building process in the form the state failure or civil war (Herzig, 2001). Therefore, it was important for Iran to counterbalance possible threats to its national and regional interests and also to take advantages of the emerged situation. Considering the fact Tehran embarked on an active and pragmatic strategy in the Caspian region. Like Turkey, Iran was among the first countries to grant diplomatic recognition to most of the new states in Central Asia following the breakup of the Soviet Union and to try to revitalize the traditional cultural and commercial ties with the region (Bahgat, 2003)[25]. However, while analyzing

---

24   In 2001, Azerbaijan bought some patrol boats from the United States and Turkmenistan bought some from Ukraine and negotiated an arms deal with Russia.

25   Moreover, official Tehran was supporting the expansion of the Economic Cooperation Organization in 1992 to include Azerbaijan, Kazakhstan, and Turkmenistan (as well as Afghanis-

Iran's Caspian policy it is possible to identify that political and commercial interests were the main drivers of its regional strategy, rather than religion. Since the early 1990s, Tehran was able to build up better relations with Christian Armenia than with the predominantly Shia Azerbaijan. Through its constructive policy in the region Iran was pursuing the role of major player in political and economic terms. During the escalation phase of the conflict between Armenia and Azerbaijan, Tehran was trying to mediate the disagreements between the two Caucasian states. However, Azerbaijan pro-Turkish policy and Iranian concerns regarding the nationalism policy within the country challenged the establishment of strong relationship between the two countries. In contrast, Armenia isolated by Azerbaijan and Turkey, moved to closer to Iran and entered into the pipeline politics of the Caspian by agreeing to the construction of a pipeline to take Iranian natural gas across its territory, which end up with the close relationship between Tehran and Yerevan (Ehteshami, 2004).

Due to limitation posed on Iran by sanctions, it was looking for another way of partnership with the littoral states. To strengthen its position in exploitation of the Caspian's hydrocarbon resources in 1998, Tehran formed a consortium with Royal Dutch/Shell and Lasmo to develop oil and gas fields on its Caspian shores (Bahgat, 2003). Moreover, Iranian companies have taken stakes in some international consortia to explore and develop oil and gas fields in Azerbaijan, Kazakhstan and Turkmenistan. In spite of slow and reluctant cooperation between Iran and littoral states, Tehran succeeded in negotiating several swap agreements with Ashgabat, Astana, and Baku during the first phase of the Caspian energy development.

4.3.5 European Union

During the first phase of the Caspian energy development, the EU had kept a low profile in the Caspian region and was not actively involved in pipeline politics as an actor. There were not defined specific EU policies and instruments for engagement in the Central Asia and South Caucasus. The main concern of the EU was political and security, which was threatened by instability and conflicts in the territories of the former Soviet states. It is worth to under-

---

tan, Kyrgyzstan, Tajikistan, and Uzbekistan) can be seen as an important step in this direction. For more details see: Bahgat, G. (2003). *American Oil Diplomacy in the Persian Gulf and the Caspian Sea.* Gainesville, Fl., USA: University Press of Florida.

line that in the 1990s the Caspian region was not considered strategically important by the European political elite (Haas, 2006). Europe's weak participation in the Caspian energy politics was explained mainly by its desire not to give up with its 'Russia first' policy and the presence of another Western actor, namely U.S., which was already actively involved in the region (Baran, 2002). However, major European oil and gas companies, such as BP, Shell, Total, Statoil and Elf were involved in exploitation of oil and gas reserves in the region. So far, to a large extent energy policy has remained within the competence of the individual EU member states foreign policies, and was a matter of national sovereignty (Haas, 2006).

Earlier in 1990s, the EU set up two programs – the Transport Corridor Europe Caucasus Central Asia (TRACECA) and the Interstate Oil and Gas Transport to Europe (INOGATE). The former aimed to develop an east-west transportation corridor from Central Asia, across the Caspian Sea, through the Caucasus, and across the Black Sea to Europe. The latter was designed to rehabilitate and modernize regional oil and gas transportation systems (Baran, 2002). Europe activated its participation in the energy projects implemented in the Caspian Basin starting form 2000, stipulated by the rise of energy security concerns and the necessity of source diversification. The change in its energy and foreign policy was underlined with the EU's Green Paper of 29 November 2000, titled "Towards a European strategy for the security of energy supply". This document mentioned the EU objectives in the field of securing energy supplies and the diversification of energy resources in order to minimize external risk factors and dependence on one source[26] (Commission of the European Union, 2000).

Intensification of the EU's interests in the Caspian region was determined by the change in its Russian politics. Rise of Russian energy dominance in the European markets and the refusal of Moscow to ratify the EU's Energy Charter Treaty, which would have given the EU access to oil and gas from Turkmenistan and Kazakhstan via the Russian pipeline network, forced Brussels to look for alternatives substitutes for Russian energy supplies and the ways of getting Caspian oil and gas reserves to the European markets. This EU's active involvement in the energy politics in the region can be seen in the second stage of

---

26   For more details please see Commission Green Paper of 29 November 2000 Towards a European strategy for the security of energy supply (COM(2000) 769 final) available at: http://eur-lex.europa.eu/legal-content/EN/TXT/?uri=URISERV:l27037

the Caspian energy development starting from 2006 with the EU report, in which the Union's main energy objectives were identified.

It is fair to conclude that immediately after the collapse of the Soviet Union in 1991, only four powers – the United States, Russia, Turkey, and Iran – had sought the ways to advance their strategic and commercial interests in the Caspian Basin. The interests of these major powers were neither mutually exclusive nor identical. The Caspian energy game can be characterized with the actors playing together and against each other, and considering it a zero-sum game. Rivalry between the U.S. and Iran left Tehran out of the energy politics, limiting Iran's involvement in the exploration and development of the Caspian resources and conducting more effective regional energy politics. On the other hand, Iran's position on delimitation of the Caspian Sea, as well as Russia's energy interests, prevented development of bigger regional energy projects during the first phase of the Caspian energy development.

4.4 Energy politics of the regional states in the 1990s

The collapse of the Soviet Union not only damaged the political system but also paralyzed the industry and aggravated the economic situation in the former member states (Bohr, 2004). Newly independent states of the Caspian region were lacking financial resources and were not able to exploit oil and gas reserves by their own capacities (Bahgat, 2003). In Azerbaijan – old oil province of the Soviets – oil production was facing reduction, because of technological constrains and lack of financial resources even before 1990s. Therefore development of the hydrocarbon resources and involvement of foreign investors had been considered by political leaders of Azerbaijan, Kazakhstan and Turkmenistan as a crucial for economic prosperity. They were ready to provide attractive terms to foreign investors, since financial resources were considered not only the source of the improvement of the economic situation in the country, but also as the mean of regime maintenance (Chufrin, 2001; Pomfret, 2005; Ostrowski, 2010)

However, the post-Soviet states were not easy places to work in and invest, due to political instability, intra and inter-state conflicts, ethnic tensions, economic and social crisis taken place in the region (Gökay, 2001). As it was mentioned before, number of political factors played much more crucial role in forming the first phase of the Caspian energy politics rather than economic interests.

Though international oil companies were ready to invest and begin exploitation works, companies and governments failed to start energy cooperation in early years of independence. This can be explained by two key factors: internal political instability and as a consequence of the former, investments risks associated with uncertainty. As directions of energy cooperation has been determined within the line of foreign policy strategies of the governments, frequent government changes and internal political wrangles among political groups suspended the work of the international oil companies for a while. Lasting struggle for the power among political elites was increasing uncertainty for investments, on other side.

In fact, political situation and events were developing differently in Azerbaijan, Kazakhstan and Turkmenistan. Various objectives and interests have stipulated the formation of the foreign policy choices of each country in the aftermath of independence.

4.4.1 Azerbaijan

Escalation of the ethnic tensions in Armenia and Nagorno Karabakh region led to the full-scale ethnic and territorial war between Armenia and Azerbaijan in 1988-1994. The war was the main challenge faced by Azerbaijan and its solution was set as key priority in nation-state building by political elite in Baku. Since territorial integrity of the country was violated as a result of the conflict with Armenia, all the political issues were approached through the prism of conflict settlement according to the principles put forward by official Baku. This also was the main principle conditioning Azerbaijan's foreign policy engagement with the regional and non-regional powers in the early years of independence. In addition, successive elite change was causing shifts in foreign policy directions and energy policy priorities till mid-1990s. Each president was trying to get international support in this question and was planning country's energy policy in the line of such partnership.

Azerbaijan's first president, Ayaz Mutalibov, who led the country for less than five months attached particular importance to the development of strategic relations with Ukraine, Iran and Russia (Bagirov, 2001). The foreign policy direction was radically changed, when Ebulfaz Elcibey came to power and prioritized development of strategic partnership with Turkey and Western countries, as a way to strengthen national independence (Bagirov, 2001). During Elchibey's ruling various interim memoranda and agreements were signed between Azerbaijani government and oil companies[27], which promised mutual economic benefits to all parties, also pursued political objectives (Baranick & Salayeva, 2005). In fact, Elchibey's government was not strong enough to prevent anti-government insurgency and ensure political stability in the country. Besides, he failed to implement economic reforms and was divested of his presidential power, first by the Parliament, and then in a country-wide referendum in 1993 (Baranick & Salayeva, 2005). Following Elchibey's resignation the cooperation with international oil companies and start of the energy projects were delayed for a while.

In summer 1993, Heydar Aliyev became the third president of the Republic of Azerbaijan. On the one hand, he managed to establish political stability, initiate economic recovery, prevent fragmentation of the state as a result of ethnic conflicts, and achieve a cease-fire agreement with Armenia (Ismailzade, 2002). On the other hand, he had reset directions of foreign and energy policy by turning the course back towards strategic partnership with Russia and Iran. Aliyev hoped that two major powers of the region would keep neutrality in settlement of the conflict between Armenia and Azerbaijan in the line of latter's position (Bagirov, 2001) However, the expectations of Aliyev's government were not met and a year later, Baku played an "oil card" and returned to the pro-Western course of the foreign policy launched under president Elcibey (Hasanov, 1998; Bagirov, 2001). In September 1994 the first oil contracts on Azeri, Chirag and Guneshli fields were signed with BP Amoco, Statoil, Turkish Petroleum Corporation (TPAO), Pennzoil, Ramco, Delta, Macdermott, Unocal and Lukoil (Bagirov, 1996). This was not only a crucial step towards improving political relations with Western countries, but also attracted huge investments from abroad to the development of the oil and gas sector. Involvement of IOCs

---

27   On 7 September 1992, an agreement was signed with the BP–Statoil consortium on the Chirag field and the Shah Deniz prospect area. The agreement gave BP–Statoil the exclusive right to prepare feasibility studies and draft contracts. On 1 October the government also signed an agreement with the Pennzoil–Ramco consortium, whereby the latter agreed to implement a $50 million gas recovery project on the Oil Rocks and Guneshli fields, where 1.5 billion cubic metres (bcm) of gas a year had been discharged into the atmosphere for many years, in exchange for the exclusive right to prepare a feasibility study for the Guneshli field. For more details see: Bagirov, S. (2001). Azerbaijan's strategic choice in the Caspian region. In G. Chufrin, *The Security of the Caspian Sea Region* (pp. 178-194). New York: Oxford University Press.

in Azerbaijan has revitalized the country's energy sector through the development of large-scale new projects and the re-establishment of existing ones.

4.4.2 Kazakhstan

On the eastern side of the Caspian Sea the situation was a bit different. Compare to other post-Soviet states Central Asian countries were less prepared to the breakup of the Soviet Union. Almost for a decade Central Asian countries were not able to set clear foreign policy directions independent from Moscow's influence. In Kazakhstan political elite was the most assiduous in trying to construct a viable successor organization to the Soviet Union (Pomfret, 2005, p. 859). Moreover, maintaining close relations with Russian Federation was set as one of the main objectives of Astana's security and economic policy (Syroezhkin, 2001; Pomfret, 2006). Therefore, foreign policy of Kazakhstan was set on pro-Russian position of Nazarbayev's government.

Although pro-Russian foreign policy was slowing and limiting entrance of foreign oil companies in Kazakhstan, the absence of the stable political system was leading to bigger uncertainty. Moreover, reforms conducted early 1990s, namely privatization policies, could not prevent severe economic recession in Kazakhstan (EBRD, 2003). The new leadership failed to build a system to ensure well-functioning market economy within the country in the aftermath of independence. Also, existence of clan culture and continuous disagreements between elites were hindering the establishment of modern political system (Schatz, 2013). The lack of sustainable political institutions and regulation mechanisms in 1990s lead to the disorganization of the government (Blanchard, 1997). Moreover, intra-elite conflict prevented realization of the initially signed energy agreements during the first half of 1990s. Despite the early Tengiz agreement with Chevron and formation of the Agip-led OKIOC consortium in 1993 to exploit offshore Caspian oil, the involvement of foreign majors in exploration and exploitation was delayed by renegotiation of agreements and by opposition from the 'oil barons' of western Kazakhstan (Pomfret, 2005, p. 864).

Development of the oil sector took a central line in government's foreign policy starting from 1995, when government launched the third stage of the privatization policy. In this stage companies were sold in part or whole, or contracted to the management of individual investors for a specified period, under an individually negotiated agreement – 'making it the most corrupt stage'

(Olcott, 2002). The period covering the second half of 1990s is associated with the wealth accumulation by the elite. In the year starting in spring 1997 a series of oil and gas contracts were signed, as the government came to an accommodation with regional barons (Pomfret, 2005, p. 866). Once, the doors have been fully opened Kazakhstan succeeded in encouraging greater foreign direct investment.

4.4.3 Turkmenistan

In Turkmenistan foreign and domestic policy were developing totally in different directions. Turkmenistan experienced short and quick transition from communism to nationalism, whereas economic liberalization was not set as a priority. The foreign policy has been built on positive neutrality and ruling elite was pursuing isolationist politics. President Niyazov was suspicious of any foreign commitments that may interfere with his power and limited the presence of the international companies in the economic and social sectors of the country (Pomfret, 2006). In the energy sector the state retained control over oil and gas resources. Foreign firms were brought in for their expertise, but in production-sharing agreements left the state in control. Such nationalistic policy was limiting and preventing international companies, especially western oil and gas companies, from starting energy cooperation with Niyazov's government.

Oil production was determined as the main priority of Caspian energy development at the end of the last century. Since, Turkmenistan does not possess significant oil reserves it received little foreign direct investment during the first phase (Shaffer, 2010). Azerbaijan and Kazakhstan became attractive to foreign investors since both were offering full legal partnership based on production sharing agreement (PSA) in the produced oil and natural gas. The most of these investments came from U.S. and European oil companies, where non-western oil companies have had little shares.

4.5 Energy projects of the first phase of development

The weakening of Moscow's political and economic influence in the region in the early years after 1991 created an opportunity for western governments and international oil companies to enter and start development of huge energy projects in the Caspian region. Despite negotiations and initial agreements, international oil companies entered into the region during the second half of the 1990s. Azerbaijan and Kazakhstan became region's main oil producers involved in energy cooperation with IOCs. Pipeline politics pursued during the first phase of the development was organized in a way that Caspian oil goes to the world markets in four directions: western, northern, eastern and southern. Leaders of Azerbaijan and Kazakhstan governments were realizing the geographical constrains of the region and the necessity of multiple pipeline policy as a tool of balanced foreign policy and economic prosperity. In fact, strategic and political considerations were the main determinants while decision-making process in the first concerning the pipeline projects during the first phase of Caspian energy development.

The first phase of Caspian energy development lead to several large-scale production projects and export pipelines. Most of the foreign investments were directed towards exploitation and development of oil fields. In contrast to oil, the natural gas production has been developed as sideline, sine gas reserves were not estimated to be huge enough and were left out of attention. Consequently, Azerbaijan and Kazakhstan were more successful in attracting massive foreign investment to development of the oil fields and got a chance to enter into the global oil market[28]. In April 1993, Kazakhstan concluded an agreement with Chevron on development of Tengiz oil field and in September 1994 Baku signed the Production Sharing Agreement with Azerbaijan International Operating Company (AIOC), an international consortium, to develop Azeri, Chiraq and Gunashli oil fields.

The main concern of oil companies at that time was how to reach the world market from the landlocked region, making the transportation question crucial. It was apparent that no single country or pipeline system could handle the volumes of oil that were to be exported from the Caspian basin (Cornell, Tsereteli, & Socor, 2005, p. 17). During the first years when oil production became revitalized, Azerbaijan was exporting its oil primarily through the northern and western routes. Baku was exporting oil to the Russian Black Sea-

---

28 Despite Turkmenistan's huge natural gas fields, it has not been so successful like Azerbaijan and Kazakhstan. Since the main focus of the international companies were oil during 1990s, Azerbaijan and Kazakhstan were much more successful in negotiating with IOCs rather than Turkmenistan and have signed contracts, which are worth billions of dollars. This also can be explained by the main focus being on oil during 1990s.

port in Novorossiysk via a northern route (Baku–Novorossiysk pipeline), to Georgian Black Sea port of Supsa through western route and small amounts of oil has been transported by rail to/through the Georgian port of Batumi. Oil export to Iran has been implemented based on swap agreement between Baku and Tehran[29]. That's why upstream investment decisions were challenged, with the fact that existent transportation of Caspian oil to world markets was possible through Black Sea transiting Bosporus. Since, number of tankers doubled in early years of 2000s and has rising huge environmental concerns, Turkish government opposed further increase of the oil transit through the already congested straits (Elkind, 2005). New pipelines were required for the shipment of the produced oil.

The turning point in this strategic pipeline game was the initiation and construction of Baku-Tbilisi-Ceyhan (BTC) pipeline, which damaged Russian politicians' dream of reviving their country's dominion in the Caspian basin and has opened opportunities for the new independent states to strengthen their sovereignties by providing access to the energy markets. Azerbaijan exports oil through western, northern and southern routes. With 1768 km in total length BTC pipeline connects oil fields in the landlocked Caspian with Turkish Seaport Ceyhan in the Mediterranean, traversing territories of Azerbaijan, Georgia and Turkey. The pipeline became operational in 2006 and ships oil from Azerbaijan, Kazakhstan and Turkmenistan. Throughput capacity of the pipeline was one million barrels per day from March 2006 to March 2009, and has been expanded to 1.2 million barrels per day by using drag reducing agents (BP, 2015).

The major development of the first phase was the construction of the new pipelines from the Caspian. The first new pipeline in the western direction was Caspian Pipeline Consortium (CPC) exporting Kazakh oil to Russia's Black Sea port of Novorossiysk. CPC was built from Kazakhstan through Russian territory, with the strong encouragement of the U.S. government and major investments by American oil companies (Shaffer, 2010). Besides, Russian and Kazakh governments and companies have been involved in the forming the consortium. Exporting oil from Novorossiysk enables companies to ship it through the Black Sea to the port of Odessa in Ukraine or to Burgas in Bulgaria, which gives an opportunity to get into Eastern European or to reach the

---

29   During the 2008 Russia–Georgia War, when the BTC was not operational, Baku increased significantly the amount of oil it exported via Iran. Shaffer, B. (2010). Caspian energy phase II: Beyond 2005. *Energy Policy , 38*, 7209–7215.

Western European/World markets by shipping oil through the Burgas-Alexandroupolis pipeline[30] (Mairet, 2006). Kazakhstan also exports oil to Russia through Atyrau–Samara pipeline system.

Through the eastern route Kazakhstan export a smaller volume of oil to China via the Atasu–Alashankou pipeline, which became operational at the end of 2005. In the eastern direction, Kazakhstan has supplied oil to Iran's Northern provinces. Tehran in turn exported additional supplies to world markets at its Persian Gulf ports, reducing transportation costs for both countries (Idan & Shaffer, 2011). Oil export in western direction implemented across the Caspian through Baku–Tbilisi–Ceyhan (BTC) pipeline. It is worth to mention that Kazakhstan also exports oil from Georgia's Batumi port, by barge across the Caspian Sea then by rail across the Caucasus.

Moreover, parallel to BTC pipeline Azerbaijan's first major natural gas pipeline the South Caucasus Pipeline (SCP), known as the Baku–Tbilisi–Erzurum pipeline has been constructed and became operational in 2006. Also, Baku supplies natural gas via the Baku–Astara pipeline to Iran's Northern provinces and began in 2006 to transit natural gas via the Iranian domestic distribution system to the Azerbaijani region of Nakhchivan (Shaffer, 2010). Realization of the first natural gas supply from the region independent form the Russian transmission system was a small success, but determining the direction of the next phase of the Caspian energy development.

Caspian energy phase one opened new opportunities for the regional states for independent development of their energy industries. First, establishment of the multiple export projects reduced dependency on existing soviet transmission network and transit vulnerability of the landlocked states. In Kazakhstan, five major oil production projects became operational and two international oil export pipelines were established, as well as additional routes for export of oil by barge and rail. In Azerbaijan, the Azeri-Chirag-Guneshli offshore multi-field oil and natural gas production project and the Shah-Deniz natural gas production project became operational. In addition, major international oil and natural gas export pipelines were established. Turkmenistan, in contrast, launched no major new production projects and received little foreign direct investment in

---

30   The Burgas–Alexandroupolis pipeline is an oil pipeline project aimed to transport Russian and Caspian oil from the Black Sea port of Burgas to the Greek Aegean port of Alexandroupolis. The priority goal of the Project is to create a new reliable and environmentally safe oil supply route for Europe, which will both help relieve the congested Bosporus and Dardanelles straits and increase European energy security. However, the project was suspended by the Bulgarian government due to environmental and supply concerns.

phase one of Caspian export. Ashkhabad's only new export infrastructure was a natural gas pipeline to Iran, which was inaugurated in 1997.

Starting from mid-2000s Turkmenistan is making efforts to make up for lost time against countries such as Kazakhstan and Azerbaijan that have received European, US and Russian backing since their independence and have set up formidable natural resources industries (Bantekas, 2011). However, continued negotiations and presence of disagreements over the delimitation of the Caspian Sea cause additional challenges for involvement of Turkmenistan in transportation projects in western direction. From 1990s till today the problem of the Caspian Sea compose the major obstacle for realization trans-regional pipeline projects passing under the seabed and used as leverage particularly by region's major powers.

4.6 Status of the Caspian Sea and regional energy projects

Though political environment was allowing and ensuring the work of the international oil companies, prospecting for new reserves under the potentially oil-rich fields in the Caspian was delayed by disagreements over delimitation of national territories in the Caspian Sea (Pomfret, 2006). In fact, with the establishment of new littoral states in the Caspian (Azerbaijan, Kazakhstan and Turkmenistan) the legal status of the sea has emerged as one of the most contentious international problems facing the region. The existence of large offshore oil and gas deposits in the area has added urgency to the need to resolve the twin issues of the legal status of the sea and the corresponding mining rights (Mehdiyoun, 2000).

Competing interests of the littoral states, intention to get more shares and maximize their profits from exploitation of the energy resources, on one hand, and unique geographical condition of the Caspian, on the other, lead to the formation of the different positions on the delimitation issue. The major powers of the region, namely Iran and Russia, in order to oppose involvement of the Western companies in the region and restrict mining rights of the other littoral states were rejecting the division of the seabed and waters into the national sectors (Mehdiyoun, 2000; Bantekas, 2011).

Iran was referring to the twin principles of consensus and condominium in the determination of the legal status in early 1990s. Official Tehran was opposing sectorial division of the seabed resources, since Iran economically is not in a position to divert scarce resources to oil exploration and production in the Caspian (Mehdiyoun, 2000). In contrast, the Russian government was not able to set clear position regarding the delimitation issue, because of internal disagreement within the government for a long time. In order to prevent development of the energy projects between international oil companies and the new littoral states, official Moscow was supporting Iran's position and suggesting the shared ownership of the sea.

Among the new independent states only Azerbaijan was active in defending its sovereignty right over the national sector in the Caspian. The government was recalling history of state practice prior to 1991[31] and referring to the rules of international law in this issue. According to official position of Baku, the waters and the seabed must be divided based on the principle of equidistant line. Such division would ensure Azerbaijan's sovereignty over its national sector by legalizing its mining right over some of the largest offshore oil and gas fields. Therefore, the Azerbaijani delegation was actively engaged in negotiations with representatives of all other littoral states.

Kazakhstan was a proponent of the Law of the Sea in determination of the legal regime. It supported the idea of establishment of internal and territorial waters and an exclusive economic zone in the Caspian Sea. In 1993, Turkmenistan passed a law declaring its jurisdiction over a 12-mile territorial sea and a maritime economic zone. However, both Central Asian states were avoiding direct contradictions with official Moscow (Chufrin, 2001).

Competing interests and different positions of the littoral states concerning the legal status of the Caspian have left the issue unresolved and ended with bilateral agreements between some of neighbor states. Turkmenistan and Kazakhstan agreed on a sectorial division of the sea through the use of equidistant lines in 1997. The agreement stated "all countries bordering the Caspian Sea must stand by the principle of dividing the water area out to a middle line until the Caspian Sea's legal status is determined" (Volovik, 2011). Similar to the agreement with Kazakhstan, Turkmenistan reached mutual understanding on basic points with Azerbaijan in the following year (Mehdiyoun, 2000). Based on mutual understanding agreed with official Ashgabat, Baku undertook a number of offshore oil and gas investments in the Caspian.

---

31   In 1970 the Soviet Union had divided the Caspian into Iranian and Soviet zones by drawing a boundary line across the sea between Astara and Husseingholi; it then further divided the Soviet sector among Azerbaijan, Russia, Kazakhstan and Turkmenistan.

However, agreements among these three littoral states on the principle of equidistant lines did not play the role of legal basis, since they could not agree on how the line should be drawn. This led to the disagreement and even to the dispute between Azerbaijan and Turkmenistan. Both disagreed on the point from which the median line should be drawn. Radical change in Ashgabat position can be explained with a desire to assume ownership over any one—or all—of the three major offshore deposits currently controlled and exploited by the Azerbaijan, namely Chirag, Kapaz and Azeri (Bantekas, 2011).

In July 1998, Russia and Kazakhstan signed an agreement on delimitation of the northern seabed, in order to delimit the northern seabed in order to facilitate the development of oil fields in their respective zones (Grau, 2001). Kazakhstan and Azerbaijan signed an agreement on delimitation of the depth of the Caspian Sea and a protocol to the agreement on 29 November 2001 and 27 February 2003, respectively.

Negotiation with the northern neighbor was developing very difficult, especially for Azerbaijan. In order to ease Russia's hard position, Azerbaijan awarded 10 percent share to Lukoil in Consortium and a multi-billion-dollar contract in 1995 and 1996 (Mehdiyoun, 2000). As Russian oil ompany became engaged in the energy development projects in Azerbaijan, Moscow proposed a hybrid plan to combine Azerbaijan's position with Iran's and Russia's positions of shared use and ownership in 1996[32].

However, the agreement between Azerbaijan and Russia was reached after 2000, when Vladimr Putin became the president of Russia. He completely changed country's position and expressed the idea to divide the sea into utilization zones, with the joint use of seabed and surface areas (Hasanli, 2010).

Russia put forward a proposal that included condominium status, 45 mile-sectors, the validity of the 1921 and 1940 treaties and application of the 1980 Convention (Hasanli, 2010). An agreement on the border depth of the Caspian Sea was signed between Kazakhstan, Russia and Azerbaijan on 14 May 2003

---

32   This development left Russia in an ironic position in 1994, while its Foreign Ministry was calling Azerbaijani oil operations in the Caspian illegal and threatening to disrupt them forcibly, its Ministry of Fuel and Power – allied with Lukoil and other powerful oil companies – was preparing to assist Azerbaijan in the same projects. The oil lobby scored a major victory in November 1994, when Prime Minister Chernomyrdin, the former head of Gazprom, met President Aliyev in Moscow and reaffirmed his acceptance of the consortium deal. For more details see Mehdiyoun, K. (2000). Ownership of Oil and Gas Resources in the Caspian Sea. *The American Journal of International Law* , *94* (1), 179-189. and Bantekas, I. (2011). Bilateral Delimitation of the Caspian Sea and the Exclusion of Third Parties. *The International Journal of Marine and Coastal Law* (26), 47–58.

(Bantekas, 2011). According to the terms and conditions of the agreements, Russia and Kazakhstan will jointly exploit the contested hydrocarbon structures 'Xvalyn', 'Centre', and 'Kurmangazy, and Russia and Azerbaijan will jointly exploit the Yalama-Samur structure (Hasanli, 2010). At the same time, all three states agreed on the junction point of the demarcation lines of the Caspian's seabed. Bilateral agreements signed between littoral states encompass sovereignty over all resources on and below the surface waters and the seabed and gave the right to these states to start energy development projects in their respective areas. However, the waters of the sea were deemed to be common property. One of the most important developments in the light of signing agreements between Azerbaijan, Kazakhstan and Russia was that these states agreed on the joint development of oil and gas fields (Bantekas, 2011).

The agreement between Kazakhstan, Azerbaijan and Russia has left Iran alone in the battle for the shared ownership in the Caspian. Consequently, Tehran had no other choice, rather than to change its official position. Iran accepted the principle of sectoral division of the seabed, but argued that such division must be equal, whereby all littoral states will get 20 percent of the share of the water and the seabed (Mehdiyoun, 2000). Indeed, none of littoral states supported this position. Since Iran possesses the smallest coastline among other counterparts and does not have huge resources in the respective area, such position gives an opportunity to challenge energy projects in the region.

Unfortunately, it was not possible to solve the problem of the legal status of the Caspian Sea during the first phase of the Caspian energy development. This problem was passed over to the second phase and became one of the main challenges in the realization of the Southern Gas Corridor, which is to be elaborated more detailed in Chapter 5.

5 The Southern Gas Corridor and Second Phase of the Caspian Energy Development

5.1 EU's supply security and Southern Gas Corridor

The idea of the southern corridor emerged during the first phase of the Caspian energy development, and has been changed several times following political and economic decisions. The main proponents of the regional transportation project were U.S. government and Turkey. The key concept behind the corridor was construction of a new transportations system from Kazakhstan and Turkmenistan, connecting with the new pipelines in Azerbaijan through the pipelines passing Caspian seabed and going in the western direction (Pflüger, 2012). The planned network would bypass Russian territory and economic independency of the regional states. As it was described in Chapter 4, Russia and Iran referring to the unresolved status of the Caspian Sea were constantly blocking the realization of huge cross-border and regional pipeline projects during the first phase of the Caspian energy development. However, at that time Azerbaijan and Kazakhstan were able to launch successful oil transportation projects without realization of the Trans-Caspian pipeline system. Kazakhstan shipped its oil to the world markets via tankers to Azerbaijan and then through the BTC line.

However, transportation of natural gas from the Central Asia was not such easy as realization of the oil export. Gas supply is by its nature generally restricted to transport via pipeline, and the supply routes are therefore rather inflexible. The situation with natural gas has been slowly changing starting from the early 2000s. Two major developments, namely discovery of the Shah Deniz gas field and beginning of the EU's diversification strategy following Russian-Ukrainian gas crisis pushed towards reopening debates and brought back discussions about the realization of the southern corridor. This time the focus was on natural gas and its transportation from the Caspian region. The growing

© Springer Fachmedien Wiesbaden GmbH, part of Springer Nature 2018
S. Amirova-Mammadova, *Pipeline Politics and Natural Gas Supply from Azerbaijan to Europe*, Energiepolitik und Klimaschutz. Energy Policy and Climate Protection, https://doi.org/10.1007/978-3-658-21006-9_5

dependency of the EU member states on natural gas imports from Russia became one of the most important issues discussed on the political agenda.

As it is mentioned in the previous chapters, there is huge difference between oil and gas markets. The shares of natural gas imports are expected to increase over the coming years. Since natural gas is less carbon-intensive than other fossil fuels, it becomes more attractive for the greenhouse gas emissions reduction strategy and climate policy. Moreover, natural gas plays an important role in the energy mix and its share expected to grow by time. Natural gas is a very flexible fuel and therefore able to respond quickly to load changes which will become more and more important as the share of volatile renewable energy sources, such as wind and solar power, in the overall energy mix increases (Devlin & Heer, 2010). Without doubt, it will require a significant extension of current gas supply systems. In contrast to oil, natural gas markets are still regionally fragmented and market transformation processes will take long time. More likely it can increase the dependency of certain consumers from one or few suppliers. Besides, natural gas trade is built on political interests and frequently affected by the changes of inter-state relations. Since the natural gas export is limited to the transmission via pipeline, it creates high dependency between upstream, midstream and downstream countries. In order to ensure energy security and avoid transit risks, both producer and consumers have to invest in new pipeline infrastructure, which takes time to plan, construct and requires significant investment upfront. The exploration and production of new gas fields and construction of pipelines is extremely expensive and time-consuming. Therefore, it is very important to have reliable partner and decide on transport route. These factors will influence and shape the dynamics of the energy and pipeline politics based on natural gas supply.

The realization of the southern gas corridor[33] became a part of Europe's diversification policy and turned to be an overarching concept at the European level. Moreover, successful realization of the initial pipeline projects, namely BTC and BTE, from Caspian Basin across the South Caucasus to Turkey intensified Europe's participation and has opened a new chapter at the region's energy politics characterized predominantly with the politicization of natural gas export. The realization of the corridor and construction of the new, long distance, large-scale cross-border pipelines from the Caspian and the Middle East developed into the "project of European interest".

---

33 The southern gas corridor sometimes has been referred as fourth corridor.

5.1.1 Europe's growing gas demand

Today, natural gas determines the European energy security strategy and it is also significant for an effective climate change policy. The new issue of climate change has increased the importance of natural gas as a relatively clean energy source, making it competitive and more attractive than other fossil fuels. In comparison with oil and coal, natural gas while burning releases only small amounts of sulfur dioxide and nitrogen oxides, and lower levels of carbon dioxide and carbon monoxide. The use of natural gas instead of other fossil fuels may thus reduce effects of the greenhouse gas emissions to the atmosphere. At the one time, natural gas will take on a greater significance in the energy mix, as it is able to support the development of renewable energy sources in the European Union as a suitable "bridging technology" (Viëtor, 2011). Since natural gas is less carbon intensive and environmentally more attractive, it can be considered as the best energy source. It can be used during the shift to the renewable energy from using coal and oil and also after the transition period as an alternative source to guarantee the energy security and successful decarbonisation strategy.

Another factor affecting the raise of the natural gas industry is the safety questions linked with nuclear energy. After the Fukushima catastrophe the reliance on the nuclear energy apparently has been undermined and speeded up "nuclear energy exit" tendency in Germany and forced the EU to undertake steps towards stricter safety regulations. If the share of the nuclear energy will decrease in the future, demand for natural gas within the energy mix will grow respectively. Hence, in the next two or three decades in the Europe's energy consumption the market share for natural gas will intensely grow and gas will dominate in the generation of electricity (Eurogas, 2008).

According to EIA estimations, in the next twenty years the natural gas consumption in Europe may grow 0.5 percent per annum on average and natural gas import may increase 1.6 percent per annum on average (Honore, 2006; IEA, 2010). In contrast, European indigenous gas production, primary in the North Sea, is expected to decline and natural gas production will not raise much above current levels in the foreseeable future (Haghighi, 2007). A CERA study shows that Europe's gas demand will increase between 2008 and 2030, to be exactly from 526 bcm/a to 622 bcm/a. These developments will lead to growth in gas imports within the EU from 232 bcm/a to 476 bcm/a in 2030. Figure 2 gives IEA projection of the growth in demand and the decline of production in the European gas market for the next two decades.

Figure 2: Natural gas production and demand (in bcm)

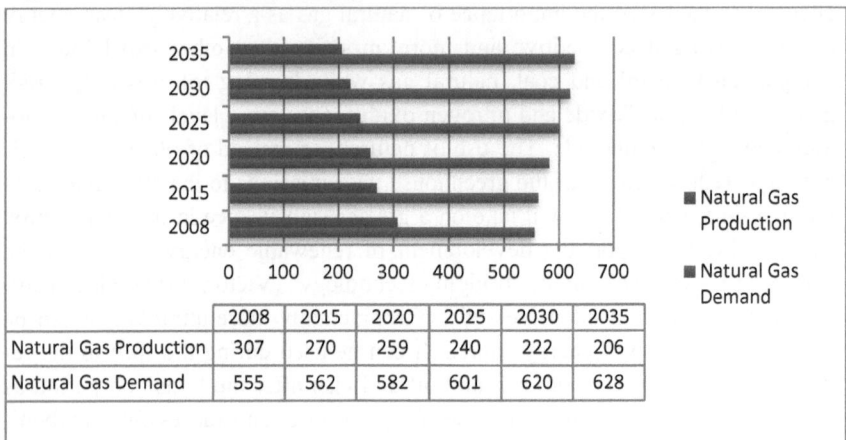

	2008	2015	2020	2025	2030	2035
Natural Gas Production	307	270	259	240	222	206
Natural Gas Demand	555	562	582	601	620	628

Source: IEA World Energy Outlook 2010 (New Policy Scenarios)

During the last decade, more than two thirds of the EU's natural gas imports came from Russia, Algeria and Norway. Some member states of the European Union are import-dependent with respect to energy. About half of the EU's primary energy needs are currently imported from outside – a share that is likely to grow to up to 70 per cent by 2030 as the EU Commission projected in 2006 (Commission of the European Communities, 2006). Due to lower cost and greater capacity, most of the natural gas is delivered to European consumers by pipelines through three large corridors: Eastern Gas Corridor – from Russia, North Gas Corridor – from Norway and Western Gas Corridor – from North Africa. Only a small amount of natural gas is imported by tankers in the form of liquefied natural gas from various producing regions.[34] A decline of indigenous gas production within the Western European countries and growing

---

34   In 2009, 33.2% of the EU-27 natural gas imports came from Russia, 28.8% from Norway, 14.7% from Algeria, 5.0% from Qatar, 3.0% from Libya, 2.4% from Trinidad and Tobago, 2.1% from Nigeria, 2.0% from Egypt and 8.8% from other third countries. Jímenez, Ana (2010) Statistical aspects of the natural gas economy in 2009 (Eurostat Data in focus 20/2010); under: http://epp.eurostat.ec.europa.eu/cache/ITY_OFFPUB/KS-QA-10-020/EN/ KS-QA-10-020-EN.PDF, p. 1.

demand for natural gas would result in increased imports and afterwards would require new gas deals outside of the EU. This will lead to new supply challenges, since it is expected that the new supplies will come mainly from the landlocked regions. According to figure 3 the pipeline transportations will have the biggest market share, despite of quickly developing LNG markets in Europe.

Figure 3: EU increase of import dependence

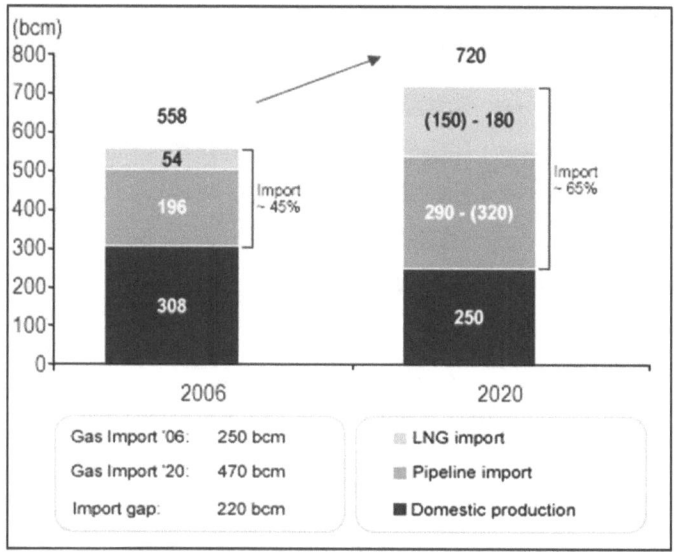

Sources: IEA World Energy Outlook; BP Statistical Review (IGI Poseidon, 2008)

To ensure energy and supply security, Europe has to diversify its supply sources and get new gas deals outside of the EU. Natural gas supply security and transit issue will remain a high priority issue in case of transportation via pipelines from the different regions.

Ukrainian-Russian disputes in 2006 and 2009 were alarm for the European Union in terms of natural gas supply security and supply diversification, emphasizing the importance of the alternative routes development. In the light of the declining domestic production and the growing energy demand, the re-escalation of Ukraine-Russia relations have propelled energy security issues back to the top of the energy and foreign policy agenda, as the gas crisis led to serious gas shortages especially in Eastern European countries resulted with human causalities and significant economic losses for Europe during the wintertime. The crisis situation demonstrated high dependency of some European countries on Russia. Although Russia provides roughly 24-25 percent of Europe's natural gas demand on average, the level of dependency on Russian gas in Central and Eastern European and Baltic countries visibly is much higher than that in the developed and well-diversified Western European countries (See Figure 4).

Figure 4: Dependence of European countries on Russian natural gas supply (%)

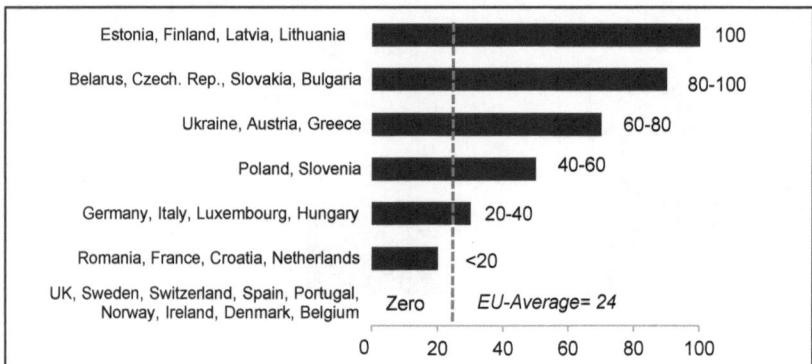

Source: Deutsche Bank Research 2014 (translated by author)

The chart demonstrates that some counties of Europe depend exclusively on Russian gas, which make them vulnerable to gas supply disruptions on the one hand, and inflexible in terms of price imposed by a monopolistic supplier, on the other hand. Moreover, lack of storage facilities and week interconnectivity between different national gas distribution systems in Europe challenge the crises management in a short-term, when disruption happens.

In order to meet the growing energy demand, the EU is strongly interested to set up the fourth corridor to enable natural gas supplies from the Middle East and Caspian region to South and Eastern Europe. Large and highly concentrated gas reserves of the Middle East and the Caspian region provide a big advantage to diversify Europe's supply routes and decrease energy dependency of the most vulnerable EU countries on a single supplier and transit country – namely from Russia and Ukraine. Realization of the Southern Gas Corridor Strategy pursued by the EU would make these large-scale reserves available for

European consumers. It would connect the EU to new sources, via new routes, and diversifying its supplier portfolio, while ensuring that, overall, gas shipments are expanded in order to meet additional future imports (Devlin & Heer, 2010). However, instability in the Middle East and North Africa has limited the ways of finding reliable and secured sources of alternative supplies from that region. Therefore, the supply of natural gas from the Caspian region, primarily from Azerbaijan and Turkmenistan, is being considered as a milestone to start the realization of the SGC.

5.1.2 Caspian gas for the corridor

In 2010 the total amount of natural gas reserves in the Caspian region was estimated around12.42 trillion cubic meters, of which Azerbaijan – 2.5 trillion cubic meters; Kazakhstan – 1.82 trillion cubic meters; and Turkmenistan – 8.1 trillion cubic meters; (BP, 2010). According to the IEA in the next twenty years the natural gas consumption in the Caspian countries will grow 1.5 percent per annum on average and natural gas production will increase 1.9 percent per annum on average (IEA, 2010). This will allow exporting large amounts of the surplus of the produced natural gas to the world markets (See Figure 5).

Figure 5: Caspian Natural Gas production and demand relation (bcm)

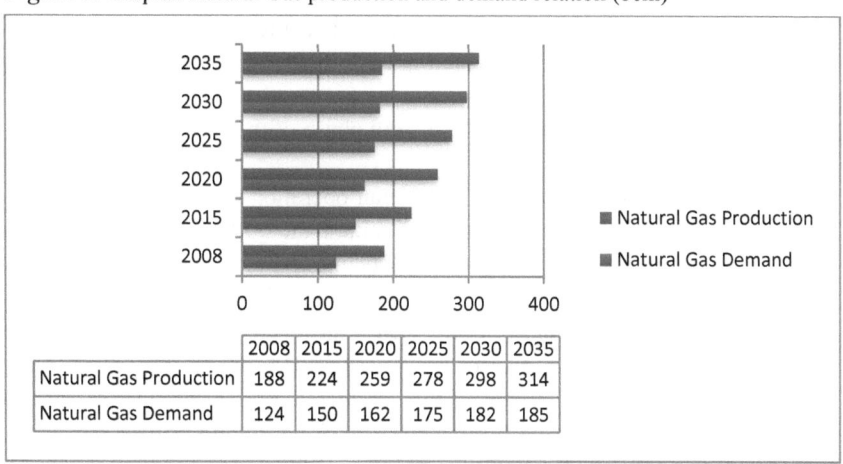

	2008	2015	2020	2025	2030	2035
Natural Gas Production	188	224	259	278	298	314
Natural Gas Demand	124	150	162	175	182	185

Source: IEA World Energy Outlook 2010 (New Policy Scenarios)

In spite of the known reserves of natural gas, the first phase of the Caspian energy development (1992-2005) was particularly build on oil production and export. This emphasized and can be explained by the commercial priorities of the companies and the difficulties to export Turkmen gas, which was seen a key natural gas country of the region. As a result of political factors and as long as, there was a reliable route of supply via Russian network Central Asian gas was not independently exported to the European markets. Since the export of natural gas from Central Asia through the Caspian Sea has been assessed politically challenging and there were not political support for it, the realization of the TCP line or in general natural gas export from the region was considered almost impossible during the 1990s.

The turning point in the energy politics in the Caspian region was the discovery of the Shah Deniz giant gas field in 1999. Located on the deep-water shelf of the Caspian Sea, 70 km south-east of Baku, in water depths ranging from 50 to 500 m, it is one of the world's largest gas-condensate fields, with over 1.2 trillion cubic meters – of gas in place, with considerable upside potential (Pflüger, 2012; BP, 2016). With the discovery of the Shah Deniz Field, the core "driver" behind the Southern Gas Corridor concept shifted from Turkmenistan to Azerbaijan. Shah Deniz provided a strong commercial driver to implement the original US-objective to develop a large-scale transportation solution to link Caspian gas to European markets without Russia (Pflüger, 2012). Today, natural gas from Shah Deniz (SD) field is expected to provide initial supply for the SGC, until other reserves become available. The exploitation of the field has been implemented in two stages, whilst the full field development constitutes the milestone of the corridor and strongly determines dynamics of the pipeline politics of the second phase of the Caspian energy development.

The Shah Deniz first stage (SD I) began operation in 2006. The field has been producing 10 bcm of natural gas per annum and 50 000 barrels of condensate per day. From 2006 approximately 6.8 bcm of natural gas has been annually delivered from Shah Deniz field to Turkey and 3.2 bcm into the local markets of Azerbaijan and Georgia. Since then it has been proved that Azerbaijan is a secured and reliable supplier of gas. During 2014, the existing Shah Deniz facilities were further de-bottlenecked which increased their production capacity from 27.3 million standard cubic meters to 29.5 million standard cubic meters of gas per day. In the first nine months of 2015, the field produced 7.2 billion standard cubic meters (bcm) of gas and 1.66 million tons (about 13.4

million barrels) of condensate, proving commerciality of the field (BP Azerbaijan, 2016).

Shah Deniz stage two (SD II) or as it called full field development aims to increase the gas production of the field and add further 16 bcm per annum to 9 bcm produced by SD I and plus 100,000 barrels of condensate, tripling overall production from the field. In fact, delivery of additional 16 bcm requires expansion of the existing SCP line and around $28 billion in capital investment (BP Azerbaijan, 2016). Additional volumes of natural gas from SD II will supply natural gas markets in Georgia, Turkey and Europe from 2018, where the share of the latter will be 10 bcm per annum. The supply from Shah Deniz field is able to ensure European energy security by brining Caspian gas to markets and also challenge Russian gas monopoly in the South Eastern European countries.

Parallel to the discoveries of the natural gas fields in the territory of Azerbaijan, two other developments have influenced the dynamics of energy politics at the beginning of the second stage of the Caspian energy development: First, it was the successful implementation of the SCP, which confirmed reliability of the trilateral strategic partnership formed by Azerbaijan, Georgia and Turkey. Moreover, planned expansion of the South Caucasus Pipeline as a part of the Shah Deniz Full Field Development project, which involves the laying of new pipeline across Azerbaijan and the construction of two new compressor stations in Georgia, will triple the gas volumes exported through the pipeline to over 20 billion cubic meters per year (BP Azerbaijan, 2016). At the Turkish-Georgian border, the pipeline will connect with other new pipeline systems providing Turkish and European markets the natural gas.

The second factor was the changing approach within the energy course of Ashgabat, which openly showed interest to join the new natural gas deal with the European Union. Turkmenistan's willingness to supply its gas to Turkish and European markets reopened discussions on the practical possibilities for the realization of the Trans-Caspian Pipeline (TCP). However, its unchanged traditional approach to pipeline politics based on "zero financial burden, hundred percent effectiveness" was causing different impediments for the materialization of TCP project *per se*. In fact, official Ashgabat more interested in exporting its energy resources to markets through the existing pipelines or where there are opportunities for expansion, like with the China route (Shiriyev, 2015). It seems that at the present time, Ashgabat is not ready to undertake financial burden of the TCP project; but it does not exclude the possibility that Turkmenistan will not join the project, when the line is finally constructed.

As long as Turkmenistan does not want to share the costs of the TCP's construction and looking for the other partners to make commitment for its realization, it seems that only natural gas from Azerbaijan will supply European market at the initial stage of the SGC. Already existing pipeline infrastructure, namely the South Caucasus Pipeline (SCP) and the rising production of natural gas following by the full development of the Shah Deniz field turn Azerbaijan to a net exporter of natural gas to Europe in the coming years.

Azerbaijan's role as a natural gas producing and exporting country has significantly increased after the discovery of the new large gas fields in its offshore areas. Each of these newly opened gas fields, Shafag, Asiman, Nakhchevan, Dan Ulduzu, Ashrafi, and Babek, has estimated volume of 200-400 bcm and according to preliminary estimates, the gas reserves in the Umid and Absheron fields are around 600-700 bcm (Rzayeva, 2010). Now, Azerbaijan's proven natural gas reserves have grown up to 2.6 tcm (Aliyev, 2012). The recent discovery of the new gas fields has tremendously shifted country's energy policy as well. It has turned Azerbaijan from oil to a natural gas producing country. It is expected that by 2025, annual production of natural gas in Azerbaijan could reach 50-55 bcm (Ismayilov, 2011).

5.1.3   EU's southern corridor strategy

The active engagement of the EU in the energy politics of the Caspian region can be observed at the second stage of the Caspian energy development. Realization of the southern gas corridor and the import of natural gas from the Caspian region (and the Middle East) turned out to be one of the main pillars of the European energy and supply diversification strategy. The importance of developing an alternative natural gas supply corridor has been described in the decisions of the European Parliament and the Council in 2006 for the first time. In 2008 the European Commission identified the "Southern Gas Corridor" as one of the essentials for the EU's energy security. Moreover, new efforts were made by the EU to develop relations and promote genuine energy partnership with the regional states, particularly with Azerbaijan, Turkmenistan, Iraq and Mashreq countries. This priority switch has been fixed in the official document of the Commission:

> A **southern gas corridor** must be developed for the supply of gas from Caspian and Middle Eastern sources, which could potentially supply a significant part of

the EU's future needs. This is one of the EU's highest energy security priorities. The Commission and Member States need to work with the countries concerned, notably with partners such as Azerbaijan and Turkmenistan, Iraq and Mashreq countries, amongst others, with the joint objective of rapidly securing firm commitments for the supply of gas and the construction of the pipelines necessary for all stages of its development. In the longer term, when political conditions permit, supplies from other countries in the region, such as Uzbekistan and Iran, should represent a further significant supply source for the EU (Commission of the European Communities, 2008).

AS indicated future energy cooperation with Uzbekistan and Iran was not excluded from the EU energy strategy and has been considered as part of the long-term scenario. Regarding the transit, the priority was given to Turkey as a key country along the supply chain. Certainly, Turkey's geographical position enables it to connect both, the Caspian region and the Middle East with the European market, underlining its geographic advantage between the two regions.

The feasibility of a block purchasing mechanism for Caspian gas ("Caspian Development Corporation") will be explored, in full respect of competition and other EU rules. Transit for the gas pipelines will need to be agreed with transit countries and notably Turkey in a way that respects both the basic principles of the EU acquis and their legitimate concern for their own energy security. The Commission will invite representatives of the countries concerned to a Ministerial level meeting to secure concrete progress and a timetable to reach agreement. It will seek to identify by mid-2009 any remaining obstacles to the completion of the project, which will be the subject of a **Communication on the Southern Gas Corridor** to the Council and Parliament (Commission of the European Communities, 2008).

Involvement and active participation of the EU Commission in the realization of the Southern Gas Corridor moved Caspian energy politics into a new stage. The EU has been supporting the major pipeline project politically and financially. During the Ministerial level meeting in 2009 held in Prague, the Southern Corridor was called as a "New Silk Road" and political support to its realization were expressed by all representatives of the countries concerned. Following Prague summit (January 2011) the European President José Manuel Barroso and the EU Energy Commissioner, Guenther Oettinger visited Baku and Ashgabat in order to get political commitment from the presidents of Azerbaijan and Turkmenistan, and to express EU's political support for the realization of the trans-Caspian gas pipeline.

On 13 January 2011, the Azerbaijani president and European Commission president signed the "Joint Declaration on the Southern Gas Corridor" and

Baku committed to export 10 bcm of gas per year to Europe. This agreement set the frame for Azerbaijan to become an important supplier of gas to the EU through the 'Southern Corridor'. In the follow-up, Turkmen president Berdimuhamedov agreed to set up an expert committee for working on technical and legal issues of gas transportation across the Caspian Sea westward, and expressed his readiness to sign a supply contract for 10 billion cubic meters (bcm) per year, if a transportation solution would be being agreed upon (Socor, 2011a).

The rationale behind the EU Commission's interest and political support for the realization of the corridor go beyond of getting just 10 bcm from Azerbaijan and Turkmenistan, respectively. The EU was targeting materialization of the alternative supply corridor with a huge pipeline capacity. The corridor was projected to become a multi-source and multi-vector network, connecting various natural gas fields with an increasingly import-dependent continent. The strategic significance of such corridor is threefold. First, it will connect the European consumers with the gas-rich Caspian region through an alternative and secured supply route. Moreover, to be designed based on scalability of the pipeline network increases its advantages, since the capacity of the corridor can be expanded when additional sources of natural gas become available. Therefore, the southern gas corridor will have the potential to supply natural gas not only from the Caspian region, but also from the Eastern Mediterranean, Iraq and Iran in the future. Second, the Southern Gas Corridor envisages a significant transit route for the natural gas supply from the Caspian Sea region and Middle East to Central and South Eastern Europe. It has a potential to diminish Gazprom's monopolistic position in the Eastern and Central European countries, which depends mostly on supplies from Russia. Furthermore, additional sources of supply available on the European markets will limit political and commercial leverages used by Moscow in these European countries. Third, ensuring natural gas supply from the energy rich regions will help the EU to meet growing energy demand in the coming decades, in the light of depletion of its local resources.

The current developments in the European energy market and the necessity of supply diversification emphasized the need for the alternatives. However, the natural gas supply from these regions to Europe is challenged by political and environmental factors, competitive nature of the supply and energy poverty driven from the lack of supply infrastructure. From the other side, there is a lack of common energy policy within the European Union. In fact, there is no agreement among the EU member states on priority or joint projects in the area of energy policy. There is no common foreign energy policy as energy policy is still determined very much by the national states themselves (Meister & Viëtor, 2011). The lack of common energy policy illustrated existing ambiguity within the European energy strategy. Some member states and energy companies were pursuing competitive policies and had different interests, which turned the European supply diversification policy into the pipeline race.

At the beginning there were proposed several transportation routes to bring natural gas from the energy reach regions to Europe. The pipeline projects proposed by various companies and consortiums to supply European natural gas markets required investments of 2–7 billion Euro (and sometimes even more) which cannot be implemented in the stages and which need to be assured of rapid build-up to high capacity utilization to ensure commercial viability (Stern, 2002). Hence, commerciality of the pipeline projects constituted an important element of the energy politics at the second stage of the Caspian energy development. Given that the willingness and capability of the consumer countries to invest in expensive new Greenfield infrastructure outside the EU and to be involved into long-term contracts based on take-or-pay principles became the major concern of the new energy game.

With the launch of the "pipeline race" political dynamics in the framework of the southern gas corridor has changed. Commercial and political interests of the different actors – state and non-state – became intertwined and directly affecting the new phase of the energy politics. Pipeline projects proposed to realize the supply form the Caspian region to Europe were targeting diverse energy markets and traversing the territories of the various transit countries. Moreover, each project intended to use different facilities through which natural gas will be delivered to the consumption markets. These factors along with commercial and political aspects were increasing or decreasing advantages of the proposed pipeline projects.

5.2 Pipeline projects of the Southern Gas Corridor – initial stage

Energy politics around the southern gas corridor became a complicated and multidimensional game with mix of political and business interests, causing frequent shifts inside of the pipeline dynamics. Number of pipeline projects was suggested to transport natural gas to European markets from the Caspian and Middle East regions as a part of the EU supported energy infrastructure initiative following Russian-Ukrainian gas disputes of 2006. At the beginning three strategic projects – the Nabucco pipeline, known as the flagship project supported by the EU Commission, pipeline interconnector between Turkey, Greece and Italy (ITGI) and the Trans Adriatic Pipeline (TAP) – competed with each other for supplies of natural gas from the energy exporting region. The Nabucco pipeline project in its initial design strongly supported and preferred by the European Commission because of its strategic advantages, the political and the economic benefits provided for upstream, midstream and downstream countries. Although the Commission often expressed its neutral attitude towards all pipeline projects, Nabucco was the EU's priority project, because of transportation route, capacity and possibility to scale up export volumes in the long term.

However, the situation is complicated with the entrance of the new projects and escalation of the political situation in the South Caucasus. On the one hand, BP, the major shareholder of Shah Deniz, announced that it is interested in the development a 1300km south-east Europe pipeline from western Turkey to Bulgaria and Romania, ending at Hungary's eastern border. Then, in order to counter the EU energy strategy, Russian Gazprom suggested construction of the pipeline network through the Black Sea to European markets called South Stream. On the other hand, on the background of the Turkish-Armenian Rapprochement Azerbaijan announced its readiness to export its natural gas to Europe by different transit routes, including LNG transportation through Black Sea.

5.2.1   The Nabucco pipeline

The idea of the Nabucco pipeline dates back to 2002, when a a small group of energy executives from Austria, Bulgaria, Hungry, Romania and Turkey sketched out a plan for a pipeline that could transport huge volumes of natural gas from the Caspian region and the Middle East across their territories into the European markets. The pipeline project was named "Nabucco", like Verdi's Opera that they attended that night in Vienna (Freifeld, 2009).

At the initial stage it was considered that the main source of supply for the new pipeline would be Azerbaijani gas and that later on reserves from Iran and northern Iraq could join. The idea of constructing this pipeline was mainly driven by commercial interests. The initial impetus was business. The Turks and the Austrians saw it as a way to get new supplies of gas from the Caspian and the Middle East – not to mention lucrative transit fees for moving it across their territories into Europe (Freifeld, 2009).

However, politics became a challenging factor for the realization of the idea. The project concept got political support from the anti-Russian politicians in Central and Eastern Europe, where the Russian natural gas monopoly was obvious and energy was uses as foreign-policy leverage. The project was seen as an opportunity to weaken Russia's influence in both regions. However, at that time major European powers[35] such as Italy, France and Germany, less dependent on Russian gas, were not very interested in supporting this project. Therefore, they were blocking any effort within the European Union to allocate funding for Nabucco, or even provide support for the pipeline a common policy until 2006 (Freifeld, 2009).

When the political approach of the western European countries changed after the Ukrainian-Russian crisis, "Nabucco" became the first pipeline project proposed for the southern corridor as a part of European gas supply diversification policy. The pipeline with proposed 31.bcm capacity and 3300 km length had to run from the Turkish-Georgian and Turkish-Iraqi borders to Baumgarten in Austria, transiting the territories of Bulgaria, Romania and Hungary. Moreover, compare to other projects Nabucco was designed upon a strong legal basis with guaranteed third party access and transmission fees fixed for a period of at least 50 years, according to the EU legislation.

---

35   Italy, under Silvio Berlusconi, and Germany, under both Schröder and his successor Angela Merkel, dragged their feet on Nabucco. France, with its nicely diversified supply of energy, had little appetite for changing the status quo. See: Freifeld, D. (2009, Aug. 24). The Great Pipeline Opera: Inside the European pipeline fantasy that became a real-life gas war with Russia.

Map 1. Nabucco pipeline

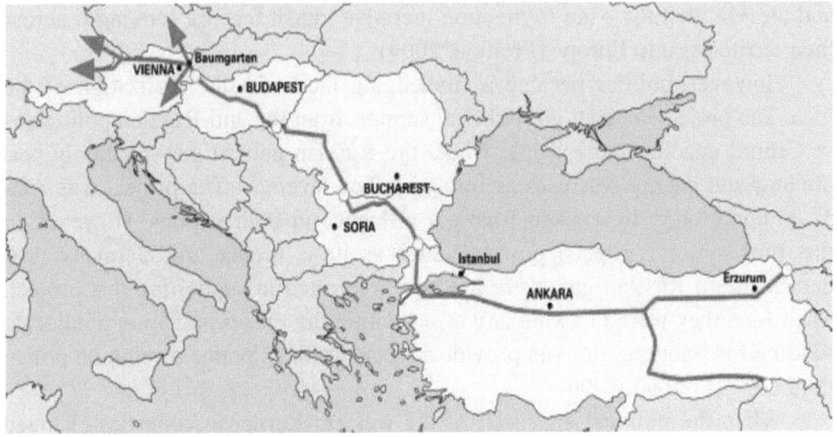

Source: *Pipeline Overview*: http://www.nabucco-pipeline.com/portal/page/portal/en/pipeline/over view

Pipeline project initiated by Austrian OMV, Turkish Botas, Hungarian MOL, Romanian TRANSGAS and Bulgarian Bulgargas had to transport natural gas from the Caspian region and the Middle East to Europe through the Balkans to Central and Eastern Europe, and in this way enable these countries to reduce their dependence on Russian gas imports. In 2008, the German RWE natural gas company joined to Nabucco Gas Pipeline International (NIC), established in 2004, with the participation of five shareholders. The Nabucco pipeline project got full political support from the EU and the concerned countries during the Budapest and Prague summits in 2009. Intergovernmental Agreements signed between the countries established the regulatory and transit framework for the project[36] (Nabucco, 2010).

Nabucco was always advertised as the Southern Energy Corridor's flagship project and as the favored option for the EU (Paul & Grgic, 2010). The project's operating philosophy was to have safe and effective transportation from inlet points to off-take points in an integrated system with upstream and

---

36 The Intergovernmental Agreement, which was signed in 2009, is valid until 2059. It is significant in the development of Nabucco as it guarantees full political support from the transit countries; defines a unique legal framework; and outlines the transport tariff methodology, - thus ensuring the stability of the project in the longer term. However, the question here could be whether it is really country issue or drops in flow of natural gas can be explained by companies policy?

downstream infrastructures (Nabucco, 2011). Also, the consortium members were responsible for the development, construction, operation and capacity trading and allocation for the Nabucco pipeline.

The strategic advantage of the pipeline project was its capacity. Due to scalability, the capacity of the pipeline could significantly increase the diversity of supply and ensure the supply security for Europe in a long term. In case of successful operation, Nabucco could transport 1,550 billion cubic meters of natural gas to the EU, over the next 50 years (Nabucco, 2011). According to RWE, Nabucco offered relatively low transportation costs on distance related tariff and as well as on wellhead-to-market costs ending with great economic savings.[37] Moreover, approved transit tariff regulations among the partner countries had to reduce the political and operational risks of the project, increasing its advantages over the other competing pipeline projects.

However, its huge capacity was seen as a disadvantage of Nabucco. Only 10 bcm of the Shah Deniz gas was committed to be supplied European markets and it could fill only one third of the Nabucco's total capacity. There was urgent need for the second reliable natural gas source to full pipeline with huge export capacity. According to the project's initial idea, it was planned to export natural gas from Iran, Qatar, Iraq, Azerbaijan and Turkmenistan (maybe also Uzbekistan) by pipelines via the Southern Corridor. In reality it is not possible to realize all these points of the project, considering the current political challenges and certain geographic constrains. Actually, not all sources that were planned to supply Nabucco were available to be delivered to the European energy markets.

The economic sanctions against Iran, because of its nuclear program, hindered the realization of the natural gas supply from its natural gas reserves. Thus, the potential Nabucco operators who counted on gas deliveries from Iran earlier announced in August 2010 that they would not build an access line to Iran, due to the current political situation (Meister & Viëtor, 2011). Besides, Qatar has been exporting a large capacity of the natural gas by LNG tankers to Eastern Asia and also to the EU. In order to connect natural gas fields in Qatar with the Southern Gas Corridor, the Iranian pipeline network would be needed. Otherwise a new pipeline through Iraq to Turkey had to be constructed, which

---

37   Nabucco's cost advantage: 41% to 73% on distance related tariff and 15% to 27% wellhead-to-market cost. This will create real economic savings of € 4.1to€9.1bnfora10 bcm/a over 25 years.   See: Dynamics of the Southern Gas Corridor: http://www.rwe.com/web/cms/en/ 257318/rwe/press-news/archive/nabucco-gas-pipeline-project/ (Retrieved: may 10, 2011)

at that time was unrealistic for both political as well as security reasons (Meister & Viëtor, 2011).

Given that, only natural gas deliveries from Turkmenistan, Northern Iraq and Azerbaijan appeared to be feasible. Since the main partners in the Nabucco project – OMV from Austria, MOL from Hungary and RWE from Germany – were involved in exploration projects in Iraq and Turkmenistan, they considered it would more likely to get additional supply for Nabucco. But in reality, the delivery from these sources were also politically challenged. In Iraq, the distribution of profits first needed to be settled between the Kurdish North and the central government in Bagdad. Furthermore, differences on the sovereignty rights of Northern Iraq between Northern Iraq and the central government, and Northern Iraq and Turkey were inhibiting gas exports.

There were also certain impediments in the case of the transportation of natural gas from Turkmenistan. Even, if Turkmenistan had shown its interest to export gas to Europe, the existing uncertainty and challenges in the construction of the Trans-Caspian-Pipeline was making the feasibility of the project less possible. Turkmenistan position of "zero financial burden" and Azerbaijan's vision of the supplies from Turkmenistan as a competing source (in a short term) moved the TCP project to second priority. What is more, due to political concerns the export of Turkmen gas through the territory of Russia and Iran was not desired. Although the project itself was strategically a great idea, its realization was almost impossible because of the changing political interests and preferences of the state actors.

Besides political complications, commercial viability of the project was under question. Uncertain price variations and rescheduling of the project implementation were undermining Nabucco's feasibility. Furthermore, Nabucco's appearance as an anti-Russian project pushed Moscow to use its political and commercial leverages for the realization of the new pipeline project called South Stream, which also got political support from some European politicians. Political support from the EU member states for the both projects was causing additional concerns for the new energy politics and leading to the open rivalry among previous partners.

5.2.2   Interconnector Turkey – Greece – Italy (ITGI)

The Interconnector Turkey-Greece-Italy (ITGI) was the second gas pipeline project named as the "project of European interest" and prioritized among other Southern Gas Corridor projects as a part of European diversification policy. The pipeline with an initial export capacity of 8-12 billion cubic meters would connect the Caspian Sea Region with Europe through Turkey, Greece and Italy. (See Map 2).

Map 2: Interconnector Turkey – Greece – Italy

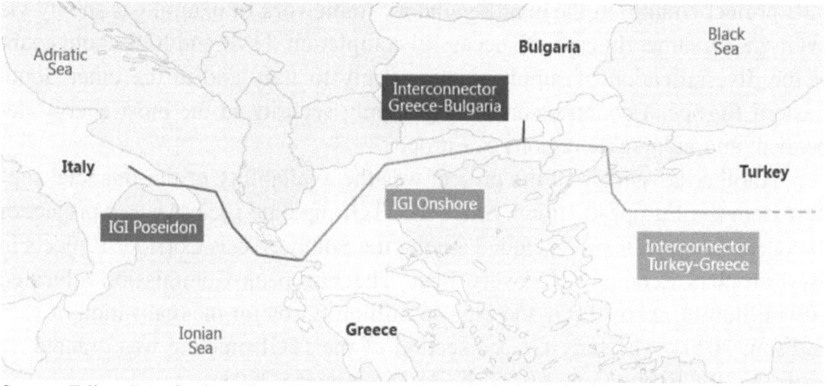

Source: Edison/Interfax http://interfaxenergy.com/gasdaily/uploads/articles/14664283408181.jpg

The pipeline infrastructure, which was planned to start operation in 2015, would consist of the already existing Turkish grid considering its limited modernization and the operational Interconnector Turkey – Greece (ITG), as well as a planned interconnector between Greece and Italy (IGI), which included two sections: IGI Onshore 600 km and IGI Poseidon 207 km, linking the Italian and Greek gas networks by crossing the Ionian Sea (Edison, 2010). The pipeline project also included construction of a new 170 km long interconnector between Greece and Bulgaria (IGB), which could transport about 3-5 billion cubic meters per year and also enter in the energy market in the Balkans

When Nabucco's realization was more-less likely, due to its capacity and doubt on availability of the additional sources, the ITGI seemed to be more realistic and viable among all the projects of the Southern Corridor (Rzayeva, 2010). This perseption was influenced by several factors. In 2007, the Protocol of Agreement between Italy and Azerbaijan supporting gas supply negotiation between State Oil Company of Azerbaijan Republic (SOCAR) and Edison for ITGI project capacity was signed. In the light of the Azerbaijan-Turkey transit gas agreement, in June 2010 Baku and Rome started intense negotiations on the transportation of Azeri gas to the Italian market through ITGI. With-in the year it was planned to begin the construction of the pipeline, when an agreement on prices reached. In addition, in 2010 the Protocol of Intent was already signed in Ankara, between the project partner companies – Turkish BOTAŞ, Italian Edisson and Greek Depa[38]. Necessary intergovernmental agreements, protocols including an agreement between Bulgaria and Greece for the implementation of IGB project composed the main regulatory framework of natural gas supply via ITGI gas pipeline. In case of successful completion, ITGI and IGB could enable the diversification of supply routes mainly to Italy and to the other South Eastern European countries enhancing supply security in the most energy dependent and vulnerable regions of Europe.

Another advantage of the project was the availability of the financial support from the European Union. Since the ITGI pipeline project was a project of European interest, it was included among the Southern Gas Corridor Projects in the European Economy Recovery Plan. The European Commission allocated 100 million Euros for ITGI and plus 45 million Euros for the realization of IGB (Edison, 2011). The Italy-Greece section of the ITGI pipeline was granted 25 years of TPA Exemption (Art.22 of EU-Directive 55/2003).

The advantages of the project were determined by the signed intergovernmental agreements and the existence of the most parts of the required infrastructure. This was cutting costs of the project and was ensuring energy security in the Southern European and Balkan countries. Furthermore, the project considered developing other regional interconnectors like Bulgaria-Romania and Bulgaria-Serbia pipeline. From the European perspective it also had to enable *solidarity mechanisms* to Greece and Southeast European countries, through reverse flow opportunities from Italy (Edison, 2011).

The ITGI was reviewed as two projects in one: the Greek onshore section and the IGI-Poseidon. These projects were lacking a credible solution to ex-

---

38   In 2005 Intergovernmental Agreement between Greece and Italy, defining the commitment of the Parties to support the realization of the Project, by means of identifying the legal framework for the offshore section and the need of securing the TPA exemption was signed. The Intergovernmental Agreement for ITGI was signed in Rome in 2007 among partner companies. For more details see: Edison. (2015, July 30). *ITGI Pipeline*. Retrieved Januray 14, 2016, from Edison Italy: http://www.edison.it/en/itgi-pipeline.

plain how more than 550 km of new-build onshore pipeline would be financed and made available on time to receive Shah Deniz II gas at the Turkish-Greek border and ship it to the Greek-Ionian coast (Pflüger, 2012). In addition, scalability constrains of the IGI-Poseidon offshore section were seen as a challenge for the ITGI to be selected by the Shah Deniz consortium.

5.2.3 Initiative of complementary projects: Nabucco and ITGI

In fact, at that time two pipeline projects called "project of European interest", namely ITGI and Nabucco, were competing for the realization of the Southern Corridor. In order to benefit from reduced tariffs and improve the chances of realizing both projects in 2010 RWE, German gas company, offered to merge two the EU backed southern corridor gas pipeline projects ITGI and Nabucco (ICIS, 2010). Following this statement there started discussions at the EU level to connect these two projects. In the first half of 2011 the European Commission started urging representatives and stakeholders of both of these projects to merge their operations to keep costs down and make the project technically and commercially viable. This was not the first time such an idea was proposed, but this push came as Azerbaijan was expected, within the next few months, to announce which supplier and project would get the rights to its Shah Deniz II natural gas field (Startfor, 2011).

At the same time, this was shedding light on some of the difficulties concerning the technical and financial aspects of both projects. In fact each project had several and specific impediments. It was politics looking for merge rather economics (Harrison & Westall, 2011). More than 20 energy companies were involved in pipeline politics and competing for the right of natural gas export. Therefore, it was believed that the idea of the merger between Nabucco and ITGI could assure that stakeholders of both projects together could realize the project more successfully. But it should be mentioned that stakeholders of the both projects were not involved in active negotiation on cooperation with each other, despite the Commission's urge.

The initiative to link two pipeline projects aimed to make IGI and Nabucco complementary projects rather than competitors. This merger would see the projects combined and built in two phases — first the "Southern Corridor Phase I" to Greece and Italy, and then a "Southern Corridor Phase II" that would spur north to Austria (Startfor, 2011). Thus would bring natural gas from the Caspian region and the Middle East to Greece and Italy through the new transportation route of Nabucco by diversifying energy supply in their energy markets. Besides, the IGI project would complete the missing part of the "South European Gas Ring".

Nabucco was facing feasibility questions, because of capacity, source and funding. The ITGI pipeline for that time being might be better positioned to receive preference from Azerbaijan for the supply of natural gas (Harrison & Westall, 2011). That is why the European Commission moved towards linking ITGI and Nabucco. Although politicians were assessing this step as strategic, it was not clear how the merge of the two projects would happen and operations regulated. From economic perspective it should be a totally different and new project. The two steps approach or division of the projects realization into two phases was implying to downscaling the southern corridor with a low level of gas flow initially. This was leading to uncertainties and raising other questions including commerciality and feasibility of the merger.

5.2.4 Trans Adriatic Pipeline (TAP)

The third proposed pipeline project of the southern corridor is the Trans-Adriatic Pipeline. Similar to ITGI pipeline project it was targeting natural gas market in Italy and the Balkan states. It was planned to transport about to 10 bcm of natural gas annually from the Shah Deniz field through Greece and Albania to Italy by crossing the Adriatic Sea, with the possibility of doubling its capacity in the future. The pipeline system will start near Kipoi on the border of Turkey and Greece, where it will connect with the Turkish network system. From there, TAP will continue onshore, crossing the entire territory of Northern Greece, its longest stretch, then onwards east to west through Albania to the Adriatic coast. The offshore section of the pipeline will begin near the Albanian city of Fier and it will traverse the Adriatic Sea to tie into Italy's gas transportation network in Southern Italy. TAP will be 878 kilometres in length (Greece 550 km; Albania 215 km; Adriatic Sea 105 km; Italy 8 km) (TAP AG, 2016).

Map 3: The Trans-Adriatic Pipeline (TAP)

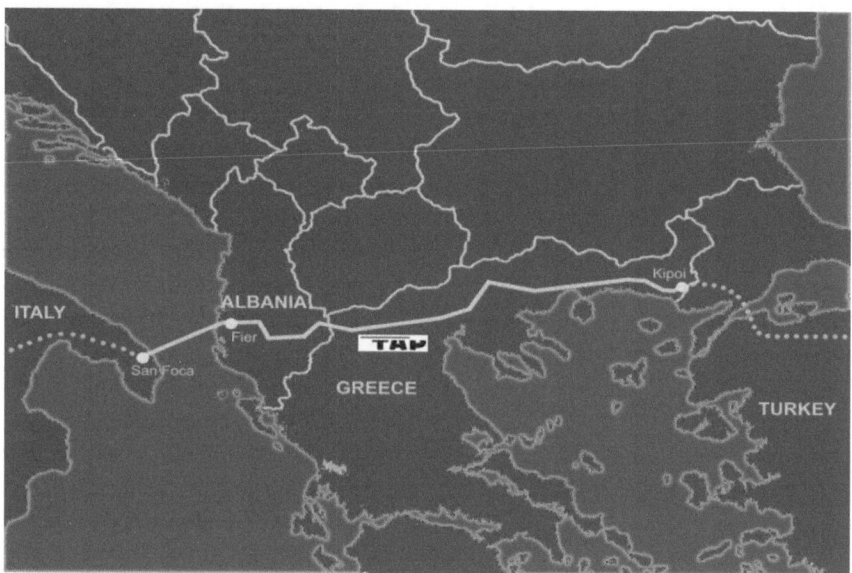

Source: TAP-AG 2016

Compare to Nabucco and ITGI, TAP has got later the status of the project of European interests. The initial shareholders of the pipeline project were Swiss EGL (42,5 %), Norwegian Statoil (42,5 %) and German E.ON Ruhrgas (15%). The chances of TAP to be selected were high, since it has Europe's most experienced and capable onshore and offshore pipeline construction management and pipeline operators, as well as credible financial support. Later new participants has entered into the project and today, TAP's shareholding is comprised of BP (20%), SOCAR (20%), Snam (20%), Fluxys (19%), Enagás (16%) and Axpo (5%) (TAP AG, 2016).

Along other pipeline projects planned within the European diversification policy the TAP is the shortest transit route. By connecting the Turkish-Greek interconnector with the Italian pipeline system, TAP was assessed as the "missing link" within the whole chain. Moreover, the construction of the additional compressors to expand the capacity of the already existing pipeline system increases commercial viability of the pipeline project. Moreover, the offshore section of the pipeline is routed through the shallowest part of the Adriatic Sea, which will ensure the long-term physical stability of the pipeline and contribute to lower gas transport costs (TAP AG, 2016). Another advantage of TAP is that it will bring cheap but high quality of natural gas to the Italian energy market[39], which is biggest consuming energy market in Europe after Germany. For Italy that plan to become an important trading hub for the European market, realization of the project is much more preferable and profitable.

TAP also targets to reach energy markets in the west Balkan countries, which are highly dependent on Russian gas, including Albania and Bosnia-Herzegovina, which are highly dependent on Russian Gazprom. The development of an underground natural gas storage facility in Albania is an attractive part of the project, which later will enhance the gas supply to the Western Balkans and South Eastern Europe. The Albanian and Greek governments called TAP the project of "national importance." Additionally, a Memorandum of Understanding and Cooperation (MOUC) signed between the Croatian gas pipeline operator Plinacro and the Bosnian gas pipeline operator BH-Gas, which increases reliability and sustainability of the TAP project.[40] On the other hand, the construction of TAP would ensure a large inflow of foreign direct investment and foster economic growth in the economically weak European countries involved in this project. TAP and the gas transported through it will contribute significantly to gas diversification in these countries and will help them develop their energy infrastructure and regional gas network connections (Kusznir, 2013).

Completed technical, environmental and economic feasibility studies and involvement of experienced companies such as Statoil, EGL and E.ON Ruhrgas at the beginning of the project into the planning and implementation of the pipeline construction strengthened the viability of the project. Furthermore, the pipeline project time planning is based on the upstream development and the pipeline will be ready to export natural gas from Shah Deniz II when production starts. However, the construction of the pipeline has been decided to be held till the commercial exploitation of Shah Deniz II has been finalized and the relevant procurement contracts have been signed between all parties.

For Shah Deniz partners, namely for BP and SOCAR, the pipeline was considered as commercially attractive because of gas market prices and possibi-

---

39   Even some describe Italian market as oversupplied, in fact, existing natural gas in the market is not "dry", namely is not in pure form. That is why the costs of final consumption are higher in Italy compare to other European countries.

40   In fact, apparent dependence of the European Union from Russian gas increases, because Balkan states almost 100 percent depend on Russian gas.

lity to reach Swiss market in the future. Besides application of TAP shareholders in 2011 to the European Commission for an exemption from the EU's Third Energy Package was promising a chance for the downstream companies to join to the project and became involved along the value chain. This was one of the advantages of the TAP project, during route decision for the natural gas transportation from the Shah Deniz field to Europe.

5.2.5 White Stream and Azerbaijan-Georgia-Romania Interconnector

There were developed two projects – one pipeline based, the second LNG based – within the southern gas corridor. These were White Stream and Azerbaijan-Georgia-Romania Interconnector. Both of the projects had presented interests of the smaller states in the Black Sea region. Indeed, it was difficult to call these projects strategically or commercially attractive. However, they played the role of leverage to influence pipeline politics within the new energy politics.

The pipeline project "White Stream", initially named GUEU (Georgia-Ukraine-European Union), was presented by Ukrainian officials during the summit-level Energy Security Conference in Vilnius on October 10-11 intended to transport Caspian gas via Georgia and the seabed of the Black Sea to Europe. Referring to the necessity of multiple pipeline export route the former Prime Minister of Ukraine Yulia Tymoshenko asked the European Union to support this project and consider it as project of European interest (Ericson, 2009). The proposed pipeline would branch off from the SCP line near Tbilisi and go through western Georgia to Supsa on the Black Sea, then continue with a seabed pipeline under the Black Sea to Crimea near Feodosia in Ukraine, linking to the Ukrainian domestic gas transit system, or, alternatively, from the Crimea to Romania's Black Sea coast, entering EU territory there. The initial capacity of the pipeline was estimated at 8 bcm, with the potential to increase up to 24-32 bcm per year (Socor, 2007).

The development of the project was planned to be implemented in three stages involving limited throughput capacities in each successive stages. For Ukraine, this could provide a substantial alternative to the Russian monopoly gas supply, and up to 40 percent of its current imports at the higher capacity (Ericson, 2009). In 2008 the Commission supported the pipeline project and it was titled as the project of common interests and further flagged as priority project (Commission of the European Communities, 2008). Despite the project attracted political support from the EU, but there were no commercial entities representing the oil and gas industry that would take the lead and implement it (Tsereteli, 2011).

There were also interests in another project envisaging a direct route from Georgia to Romania on the seabed (Socor, 2007). The idea led to the development of another project with LNG element called Azerbaijan–Georgia–Romania Interconnector – AGRI project in 2009, when the relations between Azerbaijan and Turkey was a bit escalated due to gas negotiations and Turkish-Armenian Rapprochement. Although there were already well studied and explored transportation options, Azerbaijan was looking for additional possibility of several new projects, which also would not transit Turkish territory.

The AGRI project, which involves construction of an Azerbaijani gas-processing terminal on Georgia's Black Sea coast and the transportation of gas by ship – LNG and CNG – to Romania for further shipment to Europe's domestic gas pipeline network (Pritchin, 2011). Compare to other transportation projects AGRI did not received more attention at the European level and was largely perceived as Baku's instrument to put pressure on Turkey in gas talks and against normalization with Armenia (Oxford Analytica, 2011). Azerbaijan was ready to export 8 bcm of gas on cost of 4.6 billion Euros, which questioned its commerciality. Some analysts were arguing that Azerbaijan floated the AGRI project specifically to pressure Turkey, which would not be involved as a transit state in AGRI, to get better pricing deals out of Ankara (Startfor, 2011).

In fact, development of the Liquefied Natural Gas (LNG) industry could change the dynamics not only in the European energy market but in the whole world's energy markets by establishing supply flexibility. LNG is in principle more flexible than a pipeline. Furthermore, LNG is usually not vulnerable to the transit risks, while pipelines frequently are (Luciani, 2004). However, the supply flexibility of natural gas through LNG could be very expensive and in short distance[41] from wellhead-to-market can be characterized as commercially not viable. Consequently, AGRI was entered into the pipeline as additional bargaining cheap. Azerbaijan was using the project — no matter how unrealistic and expensive it was — as a geopolitical strategy to get political and eco-

---

41   The distance for transportation of LNG should be not shorter than 3500-4000 km. Another option to export natural gas via ships in short distance can be compressed natural gas (CNG). However, in this case apart the distance, the capacity of the exported gas plays very important role. It's not commercially advantages to transport 2-3 bcm of natural gas in CNG form.

nomic leverage with all players, including the West, Russia, Turkey and Iran (Startfor, 2011).

5.2.6   South East Europe Pipeline – SEEP

At the end of September 2011 BP, the operating company of the Shah Deniz gas extraction project, announced planning a new pipeline project stretching 1,300 km across three countries – Bulgaria, Romania and Hungary – to bring gas from Azerbaijan to Europe (Blair, 2011). The new project concept was named South-East Europe Pipeline (SEEP). BP took this step with just a few days to spare before October 1, the deadline for submission of competing pipeline proposals (Socor, 2011b). The final decision on the pipeline route selected was expected to be done by January 2012.

The SEEP entered into the pipeline race with other three pre-existing pipeline projects – Nabucco, ITGI and TAP – competing for the 10 bcm of gas allocated for Europe, out of a total production of 25 bcm from the Shah Deniz full field development. According to initial proposal, SEEP would use mainly the existing pipeline networks and interconnectors. It would require laying only some 1,300 kilometers of new pipelines on several parts of the route from Central Anatolia to Central Europe. For the transportation of the natural gas from the Shah Deniz field, SEEP would use the Turkish state-owned Botas gas network for most of the distance across Turkey and existing pipelines in Bulgaria, Romania, and Hungary. From there, SEEP would use interconnectors to deliver gas to Croatia and Austria (Socor, 2011b).

The entry of the project into the pipeline race at the last minute, became very challenging for all three projects, but especially for the EU-backed Nabucco project, as SEEP looked like a radically revised and reduced version of it. The basic rationale of SEEP however was clear: it looks like a re-structured and reduced version of the Nabucco project, which dramatically avoided the investment costs that Nabucco would incur through the construction of new pipelines by utilizing, unlike Nabucco, some of the existing interconnectors in Southeastern Europe (Pflüger, 2012).

In reality, BP's SEEP was a concept rather than a full-fledged pipeline project. There were a lot of uncertainties regarding its implementation and it was not without ambiguities (Blair, 2011). However, it was clear that the SEEP sought to cut investment costs and adjust pipeline capacity to the available gas volumes committed from Azerbaijan to Europe.

Map 4: Projects of the Southern Corridor

Source: Financial Times 8 August 2011

This decision was made in order to convince partners (state and non-state) about the commerciality of such transportation, whereas realization of Nabucco was still causing doubts.

> Some observers indicated that the vagueness of SEEP's technical details were deliberate since the entire project was in effect a SD stratagem to force Nabucco's compliance to the need of SOCAR and its other SD partners that are developing in partnership with SOCAR other very promising acreage in Azerbaijan's offshore sector, like the Absheron (Total), Umid/Babek (SOCAR), SCG Deep (BP) and Shafag/Asiman (BP) fields. The potential exports of these fields should be given absolute priority over uncertain flows of natural gas that emanated from volatile areas like Kurdish Iraq and Turkmenistan were none of the abovementioned SD partners have any major upstream presence (Rzayeva & Tsakiris, 2012).

BP's objective to develop a new project concept was driven mainly by two key factors. First, the Shah Deniz shareholders had totally different cost estimates for the Nabucco project. For them, Nabucco's management underestimated the total costs of the Nabucco project, which in reality was twice more than assumed[42]. The second was the huge capacity of Nabucco, which mainly relied on accessing Turkmen gas. BP was advocating and looking for the realization of a smaller and cheaper transportation solution for Shah Deniz gas. By referring to the scalability of the capacity, BP wanted to show the advantages of the SEEP over the Nabucco project. Step by step approach pursued by the Shah Deniz shareholders aimed to adjust the capacity based on availability and size of Turkmen gas volumes via a trans-Caspian pipeline and Azerbaijan.

5.3 South Stream vs. Southern Gas Corridor

The EU-backed diversification pipeline projects are regarded by Moscow as challenging Russian energy strategy, whereas the realization of Nabucco and the Trans-Caspian projects within the framework of the southern gas corridor were considered as a threat to its interests in Central Asia and also in the European energy markets. In order to secure access to Caspian gas in May 2007 Vladimir Putin signed a declaration with the leaders of Kazakhstan, Turkmenistan and Uzbekistan to construct a Caspian pipeline with 20 bcm export capacity that would go along the Caspian coast through the territories of Turkmenistan and Kazakhstan, and up-grade the Central Asia-Centre pipeline (Feklyunina, 2008). Through enlarging energy cooperation with the Central Asian energy producers, Russia sought to strengthen its control over the energy resources and at the same time limit access of the Western countries to the energy resources of the Caspian region.

Politics of the southern corridor became intense with the announcement of the creation a joint venture between Russian Gazprom and the Italian company ENI to build the South Stream pipeline system across the Black Sea. In 2007, a memorandum of understanding was signed between the partners. Furthermore, the French EdF and German Wintershall[43] joined the project in the following years, stressing differences between the energy policy priorities of the nation states and the EU's diversification objectives.

---

42 BP came up with a cost assessment of 14 billion Euros ($19.28 billion), versus the old unrevised estimate of 7.9 billion Euros ($10.88 billion) by Nabucco consortium. See: Socor, V. (2011b, November 2). South-East Europe Pipeline: A Downsized Nabucco Proposed by BP. *Eurasia Daily Monitor, 8* (22).

43 Gazprom, ENI, EDF and Wintershall formed a consortium with the following shares. Gazprom 50 %, Eni 20%, EDF 15% and Wintershall 15%.

Map 5.   South Stream

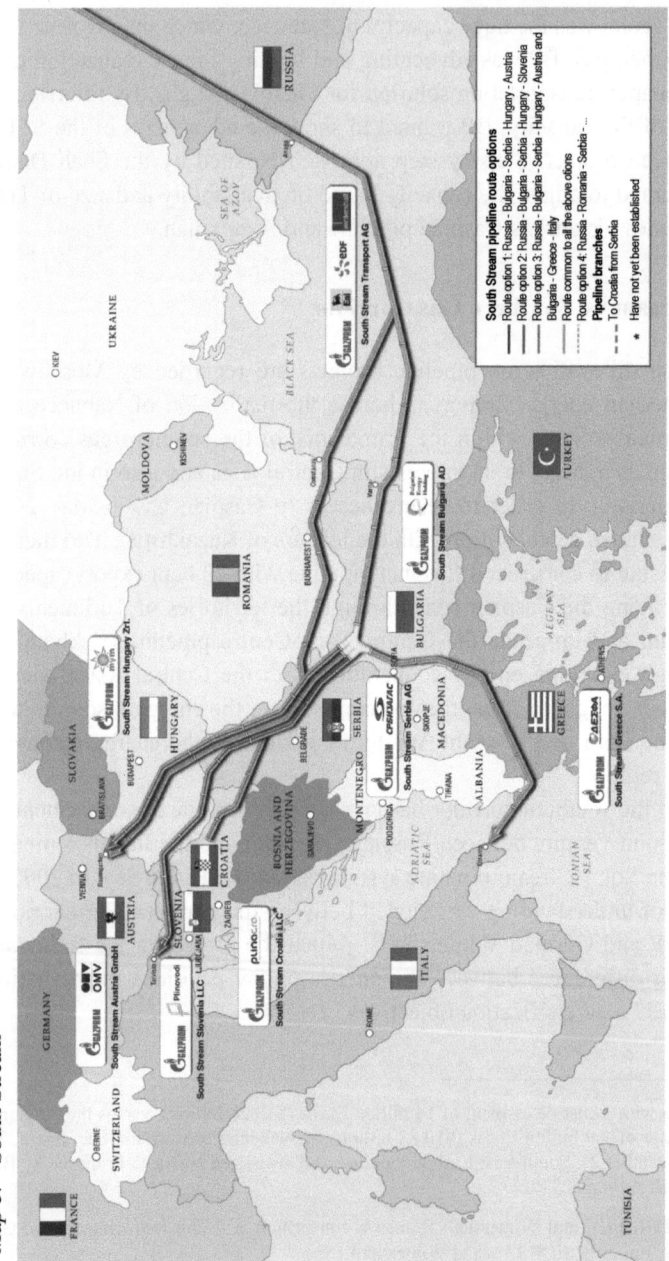

Source: Gazprom 2014

The South Stream was one of the biggest, expensive and technically challenging projects consisting of four pipeline projects, each 930 km in length to be laid from Anapa on the Russian Black Sea coast to Varna in Bulgaria in water depths of up to 2,250 meters. Originally the project was planned to be two lines with a capacity of 31 bcm/year. But, following the January 2009 Russia-Ukraine crisis, this was then expanded to four lines and 63 Bcm/year (Stern, Pirani, & Yafimava, 2015). The total cost of the project with 63 bcm capacity per year was estimated at around 40 billion dollars.

5.3.1 Understanding the rationale of South Stream

Entry of Russia with the South Stream project into the new energy and pipeline politics caused discussions around the rationale of the project and the motives behind Moscow's decision. The arguments and positions were very controversial. Some interpreted it as the Russian "bypass traditional transit countries strategy" or "transit diversification policy" already developed and advanced with the pipeline systems such as Yamal-Europe, Blue Stream and Nord Stream aiming to avoid transit through Ukraine, Belarus and Poland (Stern, 2006; Feklyunina, 2008; Ericson, 2009). Besides, for some European countries, especially for Italy, Germany and France, in this case, the problem was not dependency on Russian gas, but dependency on transit through Ukraine, which hardly can be called a reliable partner (Freifeld, 2009; Abdelal, 2010). In contrast, there were also arguments implying that the South Stream was a "bluff" designed to prevent the EU's Southern Gas Corridor (and particularly the Nabucco pipeline) from progressing (Baev & Øverland, 2010; Stern, Pirani, & Yafimava, 2015), which was considered as a serious threat to Russian energy monopoly in the European natural gas markets. As can be observed, there were two stories with different rationales.

With the realization of "Nord Stream", Moscow demonstrated its ability to break dependence on transit countries emphasizing the fact that the EU would remain as a main energy market for the Russian gas (Pflüger, 2012). Russia was and stays highly dependent on the European energy market and needs to secure the transpiration of the natural gas. Supply disruptions were regarded as financial losses and reputational damage to Russia as a gas supplier to Europe (Pirani, Stern, & Yafimava, The Russo-Ukrainian gas dispute of January 2009: a comprehensive assessment, 2009). Therefore, for Moscow it was important to eliminate transit dependence on Ukraine, rather than to worsen political relations with Europe (Stern, Pirani, & Yafimava, 2015).

South Stream has been mainly formed as part of a strategy to isolate and exert political pressure on Ukraine using gas supplies and prices (Helm, 2007). In this regard, the idea of opposition between pipeline projects, especially Nabucco and the South Stream was minimized and it was argued that the story of the South Stream began well before the Ukraine-Russia gas crisis of 2006. Additionally, Gazprom's officials stressed that the new pipeline project was not meant to outcompete the existing Southern Gas Corridor projects and would not bring new volumes of gas to the EU but merely sought to redirect already contracted volumes (Pflüger, 2012).

However, it is difficult to argue that the rationale behind the South Stream was commercial rather than political. Moscow's decision to enter pipeline politics with this huge scale project should be interpreted within the framework of 'commercial realpolitik'. Maybe on the one hand it was initiated to avoid transit through Ukraine, on the other hand, it was competing with the Nabucco targeting the same energy markets. Moreover, considering Russian dependence on the EU's energy market, South Stream also intended to undermine Europe's diversification and Russia bypass pipeline policies. Gazprom's energy strategy pursued in the European downstream countries in order to increase its export channels by means of the new pipeline projects was mainly a reaffirmation of Russia's power (Locatelli, 2008; Dasseleer, 2009).

It was more than obvious that with the South Stream project Russia was trying to put political pressure on Ukraine and also maintain its energy monopoly in Europe. Although it was announced that South Stream did not compete with Nabucco, this huge pipeline system was targeting energy markets in Hungary, Bulgaria, Serbia, Greece, Slovenia, Croatia, Austria and all these seven states signed intergovernmental agreements with Russia. Participation of the some Nabucco partner states in the South Stream project was raising more questions about the energy policy preferences within the EU. It also extended the fragmentation among the Union members.

In order to compete with Nabucco, Russia was engaged in constructive energy policy in the Central Asia and even in Azerbaijan[44] as a key regional player. Gazprom was buying and transiting huge volumes of natural gas from

---

[44] In 2009 Gazprom started negotiations with SOCAR on signing long-term contract to import gas from Shah Deniz II.

Central Asian countries. Through a policy of contracting gas from the Caspian states, Russia was attempting to control and prevent natural gas supplies to the European market from Kazakhstan and Turkmenistan (Stern, 2005; Brill Olcott, 2006). Moreover, Gazprom agreed to pay a 'European price' for the gas it imports from Central Asia, finally putting an end to the bartering system inherited from the former Soviet Union[45] (Locatelli, 2010). Certainly this was challenging the issue with getting additional sources of gas for the Nabucco pipeline and questioning the commerciality of the EU's flagship project[46].

It should be mentioned that the current pipeline dynamics were affected by the EU regulatory framework and liberalization goals. In fact, the new legal framework was causing additional challenges for the oil and the gas firms. The EU was ready to provide certain number of exemptions to its backup pipeline projects within the Southern gas corridor (Locatelli, 2010), when South Stream failed to get[47]. Somehow, it also can be interpreted as part of energy policy measures undertaken by the EU to limit actions of Russia. In fact, Europe and Russia were involved in a tough pipeline politics and diversification strategy, where each had the main objective to win and maintain its power.

5.3.2   EU regulatory framework as a challenge for South Stream

Confrontation between the EU and Russia became more obvious, while reviewing the contradictions between energy interests of Moscow and European regulatory framework. A real challenge for the Russian pipeline project or in general for its energy policy was the EU regulatory framework, including the

---

45   Throughout the 1990s, Russia purchased Turkmen gas partly through swap arrangements and partly through monetary payment. Martha Brill Olcott underlines the difficulty of estimating the real price of gas in such a system, given the opaque nature of barter transactions (Brill Olcott 2006). The new system now in place could result in substantial price increases (between 60 and 70% for Kazakh gas from 2009, according to Petroleum Economist, May 2008). Locatelli, C. (2010). Russian and Caspian Hydrocarbons: Energy Supply Stakes for the European Union. Europe-Asia Studies , 62 (6), 959–971.

46   This issue was underlined in the report of Claude Mandil, former head of the IEA. In his report on EU energy security, he pointed out that this gas pipeline could secure provision of no more than 20 bcm of natural gas per year from Azerbaijan and Turkmenistan and emphasized that: 'this is not sufficient to warrant such a huge investment. Nabucco will now only be built if it is supplied with Russian or Iranian gas or both'. Mandil, C. (2008). Sécurité énergétique et Union européenne : propositions pour la présidence française. Paris.

47   There was an argument implying that the TEP was the one of the major reasons of South Stream's fail.

Third Energy Package (TEP). TEP is set to develop a more harmonized and liberalized European internal energy market. Brussels was eager to secure its energy security and also minimize the influence of external powers on decision-making process through the political leverages. For that reason, along with the diversification policy, the EU had established a single regulatory system, comprising standards and rules, with its suppliers. Extending the Rule of Law within an essentially multilateral international system is seen as a mechanism for dealing with the issue of energy security and one that intended and would lead to the creation of a single energy market (Correlje & Van der Linde, 2006). This approach is principally promoted by the Energy Charter Treaty (ECT), which is aimed at securing international investments (Estrada, 2006) and thus guaranteeing international oil company access to producer hydrocarbon resources (Walde, 2008).

Besides, with the transit protocol, it seeks to provide third-party access (TPA) to producer pipeline networks. The TEP mandated regulated third party access to pipeline capacity based on published tariffs (or their methodologies) approved by national regulatory authorities (NRAs) as well as unbundling of transmission assets and certification of transmission system operators (TSOs) – unless an exemption from these rules is granted by an NRA and approved by the European Commission (Yafimava, 2013). Thus the TEP created major problems for Russian gas exports[48] to EU countries in terms of compliance with the changing regulatory environment both in respect of existing and new pipeline capacity (Stern, Pirani, & Yafimava, 2015). Also, the EU regulatory framework is directly conflicting with the Russia's energy policy objectives (Van Der Meulen, 2009).

In this case, two major issues have to be considered. First, Russia refused to ratify the Energy Charter Treaty, which would limit its ability to negotiate bilateral agreements with European partners in preferable way. For instance, Moscow is not ready to grant full access to its non-renewable energy resources to international firms, since the state maintains tight control over foreign in-

---

48   Gazprom has been unable to utilise full capacity of the onshore extensions of the Nord Stream pipelines – OPAL and NEL. Although the German regulator granted an exemption allowing Gazprom to use 100% of OPAL, the EC Competition Authority capped it at 50%, following which Gazprom and the EC negotiated for more than a year, and reached a solution allowing Gazprom to utilise 100% of capacity unless access requests were received by third parties (to be determined through auctions). Stern, J., Pirani, S., & Yafimava, K. (2015, January). Does the cancellation of South Stream signal a fundamental reorientation of Russian gas export policy? Oxford Energy Comment.

vestment through its increasingly strict regulation of the conditions for awarding exploration and development licenses (Kryukov & Moe, 2007). Besides, it is interested in enabling downstream access in the EU member states for Russian companies[49]. The second challenge can be determined within the framework of transit protocol and the principle of TPA. Actually, opening its pipeline network for the external suppliers will weaken Russian monopoly over the energy market and export, which is opposing its energy strategy.

In the framework of the EU's gas market liberalization policy, the profitability and feasibility of long-distance gas pipelines are in question since the rules on ownership unbundling and third party access to networks are likely to affect the investment choices of gas firms (Locatelli, 2010). Consequently, the gas firms might not be ready to commit themselves to investing in long distance gas pipelines unless at the same time they have reserved transport capacities in the pipelines that are built. In the case of ownership unbundling, since vertical integration (in this case between producer and transporter) will be legally impossible in the EU, if a supplier is not directly involved by holding a stake in the gas pipelines there would be an iterative process between allocation of transport capacity and ensuring security of supplies, which could complicate—and even considerably delay—decision making (IEA, 2008b).

Considering the above-mentioned concerns, Russia needed to apply for the right of exemption In order to maintain its market power, avoid the EU regulatory framework and realize its energy projects. However, the Russian government did not apply for the right of exemption and declared that the IGAs took precedence over the TEP and that the EC had failed to prove otherwise (Vzgliad, 2014). The Russian side appeared to believe either that the EU would be forced to agree a compromise (because of its need for the gas), or that once pipeline construction began it could be presented with a *fait accompli* (Stern, Pirani, & Yafimava, 2015).

However, in December of 2014 the Russian Gazprom announced the cancelation of the South Stream. In 2014, the estimated costs for South Stream had hit $40 billion, of which $17 billion for the onshore infrastructure in Southern Russia, $14 billion for the offshore section and $9.5 billion for the

---

49   The agreement concluded between Gazprom and BASF in 2009 provides an indication of this asset exchange strategy aimed at making Russian upstream access for international firms conditional on Russian companies being able to invest downstream in Europe. 'Downstream access' is proving to be a central element in the relations that Russia intends to develop with the EU. Locatelli, C. (2010). Russian and Caspian Hydrocarbons: Energy Supply Stakes for the European Union. *Europe-Asia Studies*, *62* (6), 959–971.

onshore European section (Franza, 2015). The decision to cancel the pipeline project was affected by following factors. First, the project became too expensive for the Gazprom, whose financial situation was increasingly precarious, owing to falling oil-indexed gas prices, the indirect effects of Western sanctions against Russia. The second factor was a drop in sales in 2014, and rising costs in other projects of the portfolio.

5.4 Economic and political dimensions of the southern corridor

There are certain risks associated with supply interdependence, which forces to undertake measures to eliminate these risks in Europe. European gas market is fragmented and market dynamics vary in South Eastern, Eastern and Central Europe. Hence, market preferences of the energy firms differ from each other and they affect the processes in the southern gas corridor. Of course, all parties involved, including state and non-state actors, had competitive political and commercial interests that were closely interlinked with each other. As it can be observed, the sketches of the pipeline projects, be it either strategic or commercially viable, had combinations of political and economic interests of certain players. Thus were formed by market rules and market dynamics. Nabucco, TAP, ITGI and SEEP separately targeted geographically vulnerable regions with a certain level of energy poverty and it emphasized the importance of the each project respectively. Proposed White Stream and AGRI projects were also targeting energy dependent regions, due to the lack of feasibility studies and existence of financial obstacles they were not considered competitive with other projects.

   Certainly, the commercial viability of a pipeline project is slightly determined by cost efficiency. Factors such as export capacity, direct link to the markets, transportation costs, transit tariffs, construction expenditures and upgrade costs of the existing supply grids play very important roles in defining the commercial advantage of the pipeline project. TAP and SEEP were seemed to be more cost effective pipeline projects, since they were developed based on a step-by-step approach and considered maximization of existing pipelines. The initial capacity of the each pipeline project was equal to the available 10 bcm of natural gas from the Shah Deniz field. ITGI and TAP projects were almost targeting similar markets and had little differences regarding the export capacity. However, technical aspects of expansion of the ITGI project and potential environmental risks associated with the transportation through the seabed via IGI Poseidon made it less competitive among other projects proposed for the southern gas corridor. The initiative with the merger of IGI with Nabucco could increase chances of the both projects, if there will be more secured sources of the natural gas for Nabucco.

Along with energy firms' market and profit interests, political considerations of the state actors were also hold up the negotiation process. As pipelines create long-term linkages and increase interdependency between several countries, the related negotiation process becomes more vulnerable in political terms. In fact, by developing new pipeline projects or considering several export routes, states are tended to use the current pipeline politics as a tool of political leverage or manipulation. However, it would be wrong to state that all projects mentioned above have been developed by political objectives. In comparison to Nabucco, TAP, ITGI and even SEEP, were almost commercial projects rather than political, whereas SEEP was a concept instead of the project.

The SEEP and TAP projects were supported and developed by the main shareholders of the Shah Deniz consortium, namely by BP and Statoil. Both energy firms were looking for a quick-fix transportation solution for 10 bcm of gas annually from SD II to Europe and their interests were limited to a "field solution," short-term and non-strategic (Socor, 2012f). Hence, through supporting the SEEP and the TAP project, BP and Statoil were involved in competition with each other over the volumes for the Shah Deniz field.

The excessive number of different states involved in these projects also requires very sophisticated and dedicated political and commercial leadership, similar to what we saw during the design and implementation of the "multiple pipeline strategy", and BTC pipeline in particular. Neither political, nor commercial conditions were present that time to demonstrate similar leadership (Tsereteli, 2011). The current energy politics in the southern gas corridor, which involve the EU member states and states of the Caspian region, as well as Russia is a complicated game where actors play simultaneously together and at the same time against each other. Even if it is not evidently visible, the Russian factor played a very important role, as well as controversial role in the European energy and diversification policy. The idea of constructing long pipeline with big capacity on higher costs from the Caspian Sea and Middle East to Europe was based on reducing the EU's import dependence from Russia and to avoid repetition of future supply interruptions similar to the Russian-Ukrainian gas crisis. Bulgaria, Romania, Hungry, Serbia, Macedonia, Croatia, Slovenia,

Austria, Italy, Greece and France became partners in the South Stream project and are willing to import more gas from Russia. The involvement of the Nabucco countries in the Russian South Stream project questions reliability of these partner countries and effectiveness of the common European energy policy.

On the other hand, it should not been forgotten that in the history of the energy supply from Russia there have never been major and frequent gas-import supply interruptions (Luciani, 2004). From previous experiences of the supply disruptions, it is possible to argue that transit countries tend to be more disruptive in energy politics rather than supplier countries. Involvement of several transit countries is more likely to threaten the supply process and cause interruptions than the dependence on just one supplier. Since natural gas is more about political concerns, reliability of partners in long-term projects turns out to be a very important issue.

6 New Geopolitics of the Southern Gas Corridor

6.1 Azerbaijan and Turkey reshaping the energy corridor

Pipeline dynamics and energy politics around the southern gas corridor became more intense and complicated, by time when the Shah Deniz consortium partners were moving close towards the final decision on the route selection. Although the European Commission backed the Nabucco pipeline and was ready to provide political and financial support for the realization of the project, in order to get direct access to Eurasian natural gas resources, the project stumbled along for a decade without real progress toward developing an adequate pipeline (Ericson, 2009). In contrast, Russia was substantially expanding its capability to deliver natural gas to Europe with the newly proposed pipeline projects. The emerged situation was forcing Azerbaijan and consortium partners to find quick and better solution to achieve their energy objectives.

In a short time, the general picture of the energy politics became more complex with the influence of the following factors. Along with the major powers, EU, Russia, U.S. and Iran, regional states, particularly Azerbaijan and Turkey, were directly involved in the decision-making process and shaping energy politics. Energy security has been reviewed in terms of national security and became a central part of the foreign policies of both regional countries. The decision-making process turned to be extremely difficult with the interests and conditions of the energy firms involved in the realization of the various production and export projects. Moreover, participation of national energy companies in the projection of the southern corridor, particularly Azerbaijani and Turkish firms, has moved pipeline politics into the new phase of the energy geopolitics. Energy firms have become actively involved in energy politics and were able to influence the decision-making process in favor of their governments. In fact, the realization of the southern corridor had great strategic importance for the all actors, state and non-state, which explicitly and implicitly were pursuing different interests. The shifts within the pipeline dynamics have been mostly deter-

© Springer Fachmedien Wiesbaden GmbH, part of Springer Nature 2018
S. Amirova-Mammadova, *Pipeline Politics and Natural Gas Supply from Azerbaijan to Europe*, Energiepolitik und Klimaschutz. Energy Policy and Climate Protection, https://doi.org/10.1007/978-3-658-21006-9_6

mined by energy policy priorities, as well as conflicting economic and political objectives of the state actors.

6.1.1 Azerbaijan's energy preferences

Since the 1990s energy politics became part of Azerbaijan's long term foreign policy and national security (Shirinov, 2011). From Azerbaijan's point of view the utilization of its hydrocarbon resources as a mean of national empowerment was always at the epicenter of the country's strategic orientation towards the Euro-Atlantic Area and also its post-Cold War security architecture (Rzayeva & Tsakiris, 2012). Following successful realization of the oil and gas projects during the first phase of the Caspian energy development Azerbaijan was able to maximize benefits from the oil export, guaranteeing the flow of petrodollars into the state budget.

The revenue from oil export ensured Azerbaijan's economic independence and sovereignty in foreign policy compare to other regional states. This was very important for the political elite in Baku, which was eager to pursue independent foreign and energy policy. According to scholars, who analyzed the political situation in the South Caucasus during post-Soviet time, economic dependence impacts state's foreign policy in a way that financially dependent countries are limited in their foreign policy decisions and in most cases do not attempt to pursue policies that contradict major powers interests (Gvalia, Siroky, Lebanidze, & Iashvili, 2013). With the oil revenues Azerbaijan has started to pursue a more active and independent energy policy with strategic goals in the second stage of the Caspian energy development. Financial independence granted with the petrodollars has enabled Baku to have direct influence in the decision-making process related to the pipeline projects within the southern gas corridor and to put its preferences as a pre-condition during the negotiations. Starting from 2005 Azerbaijan became ambitious and ready to play a decisive role in the regional and foreign policy-making. Official Baku targeted to reach wider energy markets as energy producing and exporting country. For Azerbaijan, as a natural gas producing country, securing access to the open, transparent and liberated markets, such as European gas market, as well as expanding and developing new export routes remains a priority in its energy strategy (Rzayeva, 2015).

With the southern gas initiative Azerbaijan has emerged as a one of the EU's major energy partners. In line with its new energy strategy, Azerbaijan is aiming to become an important and strategic gas exporter country in the long term and putting significant efforts into establishing a presence at every part of the value chain (Rzayeva, 2012). Therefore, selling gas from SD II on Turkish-Georgian border as a net crude exporter was not aligned with its new energy strategy. The Azerbaijani government, as an owner of the gas, did not want to transport its gas via a pipeline that belongs to a consortium that represents the interests of consumer countries, and became dependent on the infrastructure, where gas producer's interests are not represented. Natural gas is considered a strategic commodity with political and commercial advantages. Through getting a significant share along the supply chain Baku intended to get financial and geostrategic leverages as a regional power.

To achieve its energy policy objectives, the Azerbaijani government engaged the state oil company, namely SOCAR, into the energy politics. During the first phase of the Caspian energy development SOCAR was very young and weak. Foreign oil and gas companies were invited to develop hydrocarbon reserves, when Azerbaijan was highly dependent on FDI, and SOCAR had very limited technical capacity. Today, SOCAR as an international energy firm getting shares in different production and export projects, and represents Azerbaijan's interests in the southern gas corridor and other energy projects in the region and Europe. With SOCAR involved in the pipeline politics, Azerbaijan is able to protect its interests and control the potential volumes of natural gas from the wellhead till the end users over half of the value chain.

6.1.2 Turkey's energy objectives

The growing importance of natural gas, on the one hand, and the importance of the southern gas corridor for the European energy security, on the other hand, turned Turkey into the major player in the regional energy politics. Its unique geographical position between energy producing regions and the European market increase Turkey's strategic importance as an energy transit state. At present there are five active major transit pipelines going through/from Turkey: BTE from Shah Deniz (30 bcm capacity); Blue Stream from Russia (32 bcm); Iran-Turkey (1.4 bcm); Romania-Bulgaria-Turkey, supplying Russian natural gas, looping from Russian supply to Greece (17.8 bcf); and Bursa-Komotini

(Turkey-Greece), part of Turkey-Greece-Italy interconnector pipeline supply to Southern Europe (11.9 bcm) (Ericson, 2009).

Ankara prioritizes its energy interests in two particular directions. First, it has ambitions to become an energy hub not only at the regional level, but also at the international level, transforming itself into a strategic bridge between Eastern energy resources and Western markets (Bilgin, 2007). Second, it aims to secure gas for its own domestic market. Turkey's long-term energy strategy is shaped by a broad vision, taking into account the need to maintain a strict balance between its geography, foreign policy and energy demands (Akil, 2003).

Turkey's domestic energy and economic situation already plays an important role in the development of the Southern Gas Corridor. First, with the improvement and enlargement of the national transmission system it is expected that Turkey's natural gas demand will double. Hence, over the long-term Turkey's domestic natural gas market could absorb most of the gas volumes available for export from the Shah Deniz field and also some additional volumes of future generation fields in Azerbaijan. For the SD consortium partners, Turkey is a lucrative market with a high netback margin because of the short transportation distance and prices close to the European average price (Rzayeva G. , 2014).

In 2012 natural gas became the main source of the Turkey's energy. The Turkish Petroleum Corporation, BOTAS, however, holds a monopoly over the gas market in Turkey, including all natural gas transportation through the country (Ericson, 2009). The company has concluded long-term natural gas sale and purchase agreements with several states, including Russia. Russia is one of the biggest natural gas suppliers of the Turkish market and the share of the Russian gas is 60% of the total gas use. Algeria and Nigeria export LNG to Turkey, which constitutes 13% of the country's gas. It is expected that there are prospects of raise of the further natural gas supplies from the Iraq, Iran, Eastern Mediterranean, and other suppliers for the internal Turkish market. The existence of the supplier diversity will allow Ankara to negotiate gas agreements and re-export or re-sell additional volumes of gas exported from the energy producers. Therefore, Ankara is trying to increase its stake in the entire transportation chain within the Southern Gas Corridor and get greater involvement in gas supplies to Europe.

By diversifying both its natural gas suppliers and sales outlets, Turkey is insuring itself against any "hold-up" while guaranteeing a steady flow of transit revenues and substantial political influence with its neighbors (Ericson, 2009). Through increasing its shares it will get additional leverages to use during negotiations with the EU and other energy producing countries aiming to use the Turkish transmission system for the transit. It is possible to see that the Turkish preferences are not aligned with the energy security objectives of Europe. Both, Ankara and Baku aim to get maximum profit and at the same time strategic advantages from the realization of the new corridor.

6.1.3   Reshuffling pipeline dynamics

The wider picture of the pipeline politics in the southern corridor changed after inter-governmental agreement and the framework agreement on gas transit signed in Izmir between Azerbaijan and Turkey in October of 2011, which established legal and commercial terms for gas transit from Azerbaijan to Europe via Turkey and, separately, for Azerbaijani gas supplies to Turkey. The transit agreement was needed to set out specific duties and obligations on the Turkish side to secure transiting Shah Deniz II gas to the Turkish-European border (Rzayeva, 2015). At this stage, Turkey and Azerbaijan entered the pipeline politics as key decision-makers, with the power to direct to rules of the game. Transit agreement was a step forward in lasting and complicated gas negotiations between two partner states.

6.1.3.1   Transit negotiations

As history shows, there are specific challenges and success stories related to transit pipelines. If a transit state is dependent on foreign development investment and also is off-taker from the line, like in the case of Georgia, it will be less interested in supply disruption. The dynamics of transit pipelines made the concept of *obsolescing bargain* an important issue to be considered as a threat to security of supply. Since transit pipelines once built and start to operate, become vulnerable to the obsolescing bargain by putting transit state in more favorable position by shifting bargaining powers. As Vernon describes, "almost from the moment that signature dried on the document, powerful forces go to work that renders the agreement obsolete in the eyes of the host government" (Vernon, 1971). In this case, obsolescing may take the form of re-negotiation of transit terms and change in payment procedure. Negotiation over transit terms of Shah Denis phase one between Turkey and Azerbaijan is an interesting example for that sort.

Shortly after the discovery of Shah Deniz field, Baku and Ankara signed a purchase and sale agreement for the delivery of 6.6. bcm of natural gas per year to Turkey via South Caucasus Pipeline (SCP) starting form 2007. According to the gas agreement signed in 2001 Turkey had to pay low price with respect to high transit fee for the natural gas from the first phase of Shah Deniz production. In order to make this low price more acceptable for Baku, it was agreed that it would be renegotiated one year after the start of the gas deliveries to Turkey (Lussac, 2010). For Baku the prize would be won from re-negotiating a transit agreement had three dimensions: acceptable transit fee, relatively fair price for sold natural gas in the territory of Turkey; and access to other European markets through Turkey.

Ankara had its own interests in this energy game, which cast a shadow on the gas agreement between Baku and Ankara (Pritchin, 2011). Re-negotiations started in 2008, but reached a deadlock, since both sides had different positions concerning transit terms. Turkey was willing neither to pay more for the Azerbaijani gas nor get agreed on decreasing transit fee. Moreover, Ankara expressed its intention to buy and resell gas from Shah Deniz field in the European markets. On the background of the growing importance of the southern gas corridor, Turkey wanted to become an energy hub for the EU and aspired to be either the owner of transit gas or to easily obtain 15 percent of the fuel volume for transportation (Pritchin, 2011). This was an unacceptable deal for Baku and for the EU.

Turkey was aware of its importance as a key transit state within the supply chain and was manipulating with its geographical position. However, Ankara was not ready to miss the opportunity and advantages from the realization of the Southern Gas Corridor. If all parties feel they are benefiting from the project, they will have an incentive to stay with it and to work out any conflicts or disputes that may arise (ESMAP, 2003). After two years of negotiations both parties were able to get agreed over the new transit terms. Since the Turkish energy market is a major consumer of Azerbaijani gas and Azerbaijan is a key energy supplying country within the southern corridor initiative, both parties need each other to implement their commercial interests and achievetheir policy objectives. By agreeing on new transit terms, Azerbaijan and Turkey solved the problems related to transit fee, gas price and volume of natural gas, supplied from Shah Deniz phase I.

However, the main steps toward realization of natural gas supply from Azerbaijan to European markets were signing of another gas agreement between Turkey and Azerbaijan on Shah Deniz phase II. It was believed that after signing this agreement the dilemma around Caspian energy could be easily solved. The second phase of negotiations were even more tough and difficult. Hence, prolonged negotiations over the new gas agreement between the two states led to growing uncertainty in implementation of the whole east-west supply chain.

More than a year both sides were negotiating over the terms of the new gas contracts. Both parties had different positions concerning the supply volume and transit terms. The setting of transit terms for a long period has always been a difficult and controversial issue. Since there is no 'objective' or 'fair' way to set transit fees, the outcome, in the form of the transit agreement, depends upon relative bargaining power and the skill with which that power is used in the negotiations between the transit government and the transit pipeline company (Stevens, 2009).

6.1.3.2 Izmir Agreements of 2011

Energy cooperation between Azerbaijan and Turkey entered into a new phase in 2007, when SCP line became operational. Azerbaijan was exporting 6.3 bcm of natural gas annually from SD I to Turkey. The initial gas prices[50] were set bilaterally, using oil-linked formula and revised upwards as oil prices rose (Pirani, 2012). In comparison with gas prices those Turkey was paying for Russian and Iranian imports, import prices of Azeri gas were lower. However, Azeri side was not happy with the situation.

In 2009, Azerbaijan and Turkey started lengthy intergovernmental negotiations over transit and setting new base price for imported natural, and for Turkey to pay retroactively at the new, higher price levels for gas purchased in 2008-09. The process took two years and was challenged with political and commercial factors. The Turkish-Armenian Rapprochement process of 2010 had negative impacts on the negotiations between the two partner states and

---

[50] The base price used from 2007 was widely reported to be $120/mcm.

prolonged the achievement of the agreement. In October 2011 Azerbaijan and Turkey signed the so called Izmir agreements, determining the conditions and terms for transiting Shah Deniz gas.

Izmir Agreements were the end of the lengthy negotiations on the transit and the price setting, which was one of the main issues needed to be solved before the final investment decision on Shah Deniz II and the selection of the transportation route in early 2012. Furthermore, the Izmir Agreements removed existing legal and commercial obstacles to gas transit from Azerbaijan to the EU via Turkey, making it possible for suppliers and consumers to enter into commercial transactions directly.

The transit agreements envisaged two possible options for the transit of Azerbaijani gas to Europe through the Turkish territory. One option was to use The BOTAS-operated pipeline system, conditional on certain upgrading for this purpose. Initially the use of Botas's existing transport system for transportation of Azerbaijani gas from SD II was promoted and supported by BP within its SEEP project. The rationale behind this was avoiding additional costs, keeping the project less expensive instead building a new standalone pipeline. However, the Turkish grid, even upgraded will have little spare capacity especially east of Ankara for transporting bigger natural gas volumes coming from Azerbaijan and other sources (Rzayeva G. , 2014). It was obvious that Shah Deniz consortium partners will be not able to use the existing Turkish transmission system, and there was a need for a new pipeline.

The second option was having a new transit pipeline, which has been identified as a trans-Anatolian gas pipeline, to be jointly built across Turkey (Gültekin Punsmann, 2012). From the beginning, the Azerbaijani side made clear that it would prefer building a new pipeline, namely Trans-Anatolian pipeline (TANAP), in order to secure its interests in controlling the potential volume of gas from the wellhead till the end users over half of the value chain (Rzayeva G. , 2012). With the new pipeline higer volumes of gas from other sources would be possible to export. Moreover, Shah Deniz consortium shareholder was supporting the official Baku position on this strategic decision to build a pipeline with the scalable capacity through Turkey[51]. The agreements

---

51   In Izmir, along with state representatives Shah Deniz consortium also was presented. Botas
      CEO Fazil Sener and Rashid Javanshir, head of BP's operations in Azerbaijan, signed a tran-
      sit agreement for Shah Deniz gas from the Georgia-Turkey border to the Turkey-Greece and
      Turkey-Bulgaria borders. Javanshir acted on behalf of the Shah Deniz producers' consortium,
      in which BP holds the operating rights. See: Socor, V. (2011c, November 1). Azerbaijan And

had direct impacts on the pipeline competition by changing the dynamics and adding new actors to the game.

Following the Izmir Agreements on December 26, 2011 Azerbaijan and Turkey signed a Memorandum of Understanding to build a Trans-Anatolia Gas Pipeline (TANAP) to Europe. TANAP became an inevitable game changer for the entire southern gas corridor by replacing Nabucco East through the Turkish territory and determining the future of the natural gas supply to Europe from the Caspian Basin. The decision on construction of the new pipeline via Turkey to Europe underlined the differences between strategic and economic preferences of Europe, on one side and Turkey and Azerbaijan, on the other. In reality, Baku needed a new export route for its future gas production, whereas Ankara acquired additional infrastructure for the alternative sources of fuel. The evolution of the new plans has been notable for the way that both Azerbaijan and Turkey have sought to take a greater share of the control of transport and marketing arrangements (Pirani, 2012). Here, along Azerbaijan and Turkey, the interests of the companies involved were playing a decisive role and influencing the pipeline dynamics per se.

6.2 Trans-Anatolia Pipeline – TANAP

TANAP has been considered as Azerbaijan's genuine pipeline project in the southern corridor, and became its direct road to Europe. Cooperation between the two partner countries, Turkey and Azerbaijan, in the energy sphere can be described as a response to Brussels' energy policy, which failed to arrive at coordinated decisions and implementation of the Nabucco project (Zhiltsov, 2014). The pipeline project got the same strategic importance for official Baku as Baku-Tbilisi-Ceyhan oil pipeline. The key advantage of the project is its scalability. The pipeline is designed to be scalable with a 56" diameter (Rzayeva G. , 2014). At the beginning it will transport SD II gas, and then additional volumes of natural gas from other fields, when Azerbaijan doubles gas production[52]. The TANAP project will be connected to the expanded SCP line

---

Its Gas Consortium Partners Sign Agreements With Turkey. *EDM.* and Rzayeva, G. (2012). A Complicated Corridor: Gas to Europe - it's not just economics. *Caucasus International , 2* (2), 141-159.

52   Azerbaijan expects to double its current gas production to 50-65 Bcm/y between 2025 and 2030.

on the Turkish-Georgian border, run from there to the Turkish-Greek border and will then connect with the European supply network, constituting an integrated transmission system[53]. The total length of the new pipeline will be 1,541 km, with the initial capacity of 16 bcm per year in the first stage of operation. The total transportation capacity of the line is about to be increased in the second stage of operation up to 23 bcm per year in 2023 and to a capacity of 31 bcm per year by 2026, with the construction of 7-8 high-pressure compressor stations (Rzayeva, 2014).

Map 6: The Trans-Anatolia Pipeline (TANAP)

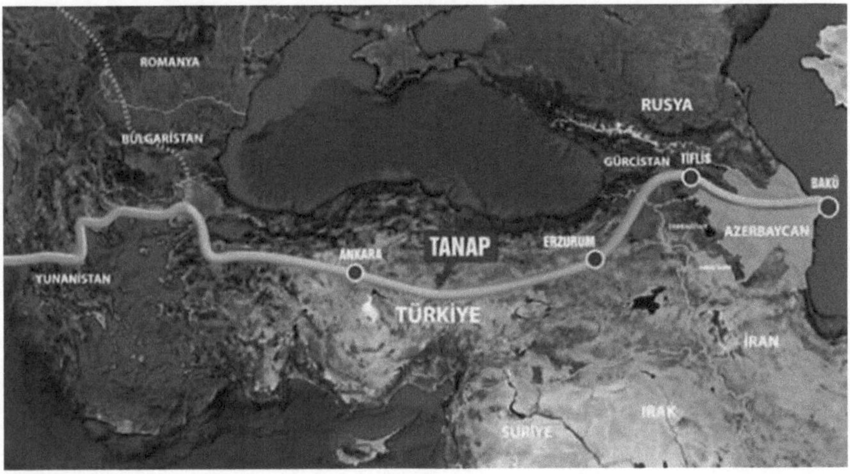

Source: TANAP 2014

The initial shareholders of the new pipeline project were SOCAR (80%), BOTAS (10%) and TPAO (10 %). However, official Baku has made it clear that it would prefer to involve SD II partners, namely BP, Total and Statoil in TANAP in order to ensure its energy security objectives (Hulbert, 2012). In 2012, it was reported that SOCAR had agreed to bring the other Shah Deniz consortium members into the TANAP consortium, by selling 12% to BP, 12% to Statoil and 5% to TOTAL (Rzayeva & Tsakiris, 2012b). Nevertheless, SOCAR was

---

53   Before the final decision was made on the transportation route, it was planned that TANAP
      will run till the Turkish-Bulgarian border. When TAP project has been selected as a main
      route for transportation of SDII gas, consortium members revised the project.

interested to maintain the majority of the shares. On the other hand, the SD II partners insisted on veto rights over the decisions SOCAR will take about technical and financial issues as project operator (Rzayeva G. , 2014). Negotiation between the partner companies and the state company was very tough, and SOCAR refused to grant the veto right to the SD II partners. After the negotiation only BP has joined the project, whereas Statoil and Total refused to do so, without having veto right. Their shares were distributed among the shareholders in the project, namely SOCAR, BP and two Turkish companies (Chazan, 2013). Following the process of acquisition the shareholding percentages on TANAP project as follow were: SOCAR 58%, BOTAS 30% and BP 12% (Kok & Dag, 2015).

TANAP is a project that suits the energy and foreign policy interests of Azerbaijan and Turkey in particular. The projection of TANAP has influenced the pipeline dynamics within the whole supply chain. In Azerbaijan's case, this involvement is seen as a first step in raising Azerbaijan's economic and strategic influence, with further expansion of gas output from fields other than Shah Deniz being linked to further such steps (Pirani, 2012). In a wider perspective, TANAP is a game-changer, with multiple ramifications across the space from Ashgabat and Baku to Vienna and Brussels (Socor, 2012a). With the emergence of the TANAP pipeline project in the southern gas corridor, the level of uncertainty around the Caspian gas politics has decreased, and at the same time it has shifted the power relation between different players. TANAP thus has had different implications on the initial projects of the southern gas corridor and the Trans-Caspian Pipeline.

6.2.1 Implications of TANAP for the regional states

The failure of Brussels' pipeline policy to come forward with the strategic pipeline project designed to bring huge gas volumes to Europe bypassing Russia, created an opportunity for Azerbaijan and Turkey to enter pipeline race with their TANAP project. The new pipeline project was approved by the EU and regarded as a foreign policy victory of Azerbaijan, which needs more transportation facilities to sell the gas produced within the second phase of the development of Shah Deniz and beyond (Zhiltsov, 2014). From Azerbaijan's standpoint, the pipeline project was an optimal solution, as Nabucco consortium had never resolved the issue of accepting Azerbaijan's State Oil Company, SO-CAR, as a partner in that pipeline project (Gültekin Punsmann, 2012). With the realization of the TANAP project, Azerbaijan will maintain control over the transportation and can sell its gas directly to European consumers on the Turkish border based on European gas market price, which will ensure flow of revenues from the direct gas trade. In addition, since Azerbaijan holds a controlling position in the transit project, it has got a decisive role on decision concerning the selecting the pipeline project or transportation route for SD gas. Strategic importance of TANAP project equals to the BTC line for Azerbaijan's national development.

The trans-Anatolia project confirmed Turkey's role as an energy corridor to Europe and intersection of multiple supply routes for Turkey itself (Socor, 2012a; Gültekin Punsmann, 2012). The advantages of construction new transmission system are twofold for Turkey. First, Ankara can easily reach and meet growing natural gas demand in the western industrial cities of Turkey[54]. Second, the launch of the Azerbaijan-Turkey pipeline project constitutes a leverage, which Ankara can use during negotiations with Moscow, to soften the terms of gas supply contracts. With the new pipeline project Ankara along with Baku will get control over the transit on natural gas from Azerbaijan through Turkey to Europe and play a crucial role in the European energy security framework.

Another regional country that directly will benefit from the transit pipeline project is Georgia. The upgrade and extension of the SCP[55] line will doubling the capacity of the pipeline going through Georgia and at the same time increase the revenues from the transit.

Although Azerbaijan envisions itself as natural gas producing and major transit country for Central Asian gas, the trans-Anatolia pipeline does not envisage a transportation solution for Turkmen gas to Europe (Socor, 2012a), as the transportation capacity of the new pipeline project is limited to some certain volumes. For the moment of the project agreement the priority was to guarantee

---

54   Turkey is committed to importing 6.6 bcm from Azerbaijan annually until 2017, and 6 bcm annually afterward. Azerbaijani gas enters from Georgia at Erzurum in Turkey's east, but Turkish gas demand is concentrated at the opposite end of the country in the west. Socor, V. (2012a, January 5). *Trans-Anatolia Gas Pipeline: Wider Implications of Azerbaijan's Project*. Retrieved January 12, 2012, from Jamestown Foundation: http://www.jamestown. org/single/?no_cache=1&tx_ttnews%5Btt_news%5D=38846&tx_ttnews%5BbackPid%5D=7 &cHash=dc04cb9a31540c9f38bd052aac6cd360

55   Known as Baku-Tbilisi-Erzurum or South Caucasus Pipeline (SCP), the line has a declared capacity of 8 bcm per year at the Georgia-Turkey border, but has been operating invariably below capacity.

the supply of natural gas from SD II and further development of Azeri gas fields. The link-up with Turkmenistan has been put in a follow-up stage by Azerbaijan government, since the realization and timing of TCP line was unsure. Moreover, the decision on capacity of the pipeline was influenced by SD consortium partners' positions, which had different approach than Azerbaijani government representatives concerning the energy politics and natural gas transportation from the region.

6.2.2 Implications of TANAP for the projects of the Southern Gas Corridor

TANAP became a game-changer for the European pipeline politics and for the whole southern corridor initiative. Three European pipeline projects, Nabucco, the ITGI and TAP, were competing for 10 bcm of SD II gas, where the European Commission has clearly prioritized the strategic importance of Nabucco over the non-strategic ITGI and TAP. The Turkish-Azerbaijani decision to construct TANAP reshuffled the dynamics of the pipeline dynamics in the southern corridor, whilst Nabucco had finally lost credibility in the form proposed[56] (Socor, 2012d). Given this change, the EU announced supporting the southern gas corridor as an overall concept, however, prioritizing supply diversification for countries over-dependent on the Russian Gazprom. On the other side, Azerbaijan and the other Shah Deniz consortium partners were favoring and looking for more doable, smaller and cheaper projects, namely SEEP, ITGI and TAP, rather than the EU's strategic project (Socor, 2012c).

Indeed, TANAP was requiring certain modifications of Nabucco pipeline project. In fact, Nabucco was facing commercial challenges and project launch was delayed as a result of the European financial crisis (Socor, 2012a). After BP's proposal of SEEP aiming to use existing pipeline network from Turkey to Europe, TANAP was the second Nabucco-substituting initiative. In fact it did not target cancelation of the whole pipeline project. TANAP has replaced the Turkish section of the EU-backed Nabucco pipeline and lead to the downsizing of the initial Nabucco project requiring construction of a shorter pipeline from the Turkish-Bulgarian border to Vienna. Theoretically it did not "kill" the Nabucco project per se, but, it gave Nabucco another lease on life and a new role by reducing its length and costs. A shorter and cheaper Nabucco would

---

56   31 bcm of gas annually, starting from Turkey's east, its €8 billion ($10.6 billion) cost an underestimate, and being unbankable in Europe

still need to become "bankable", eligible for loans, to finance its construction (Socor, 2012a).

As expected, with the entrance of the TANAP project the Nabucco consortium reconfigured its project and introduced "Nabucco-West", which would link up with TANAP at the Turkish-Bulgarian border and continue to Baumgarten in Austria, for a capacity half as the originally designed 31 bcm per year. The new pipeline would run a distance of 1,300 kilometers, instead of 3,900 km of the old Nabucco. Nabucco-West envisaged a diameter of 1,200 millimeters (48 inches) along the route, and an initial capacity of 10 billion cubic meters (bcm) annually, scalable to 23 bcm annually in the second stage, contingent on additional Caspian volumes. The new size would lack spare capacity for future volumes of natural gas from Central Asia, particularly from Turkmenistan (Socor, 2012d). This was conflicting with the EU's energy policy objectives. However, TANAP has got the support not only from Shah Deniz consortium partner, but also from the EU, US and energy firms involved in the pipeline politics in the southern gas corridor.

6.3 Energy firms and their interests

Over the past decade firms individually and at the same time, with the support of their governments have become active players in international scene by operating in the foreign markets. Consequently, they are directly involved in developing business and diplomatic relations with the host governments, which also engage the firms in geopolitics in the certain regions. Through the so-called business diplomacy firms not only pursue fulfillment of their commercial interests, but also act as political agents of their governments. With other words, firms operating and investing in different host countries constitute the means of soft power[57]. Given the importance of the energy sector, the role that energy firms play in shaping energy politics and decision-making processes have to be considered as significant factor. Energy firms are interested in maximizing revenues and returns, and to secure greater long-term political and strategic influence in the regions involved.

---

57   Soft power is the ability to affect others to obtain the outcomes one wants through attraction rather than coercion or payment. Moreover, soft power rests on the ability to shape the preferences of others. For more see: Nye, J. (2008). Public Diplomacy and Soft Power. *ANNALS of the American Academy of Political and Social Science , 616* (1), 94-109.

Several international energy firms have started operation in Azerbaijan since 1994, when the "Contract of the Century" was signed. International involvement in the country's energy sector and financial support revamped technical infrastructure in Azerbaijan's oil and gas industry. BP was one of the biggest energy companies involved in development of the both energy sectors, whereas national oil company was too weak to launch exploration and energy production projects independently. In 1996 seven oil companies signed Production Sharing Agreement (PSA) with the Azerbaijani government on development of the Shah Deniz field, where BP and Statoil became the biggest shareholders within the consortium[58] each with 25.5% of shares respectively. BP was appointed operator and Statoil Chairman of the Shah Deniz Gas Commercial Committee by the consortium partners[59]. Shah Deniz partners were providing technical facilities and invested in the project development. Therefore, they were more influential during the negotiations and decision-making process concerning the field development and setting exportation conditions.

Despite commercial interests of the companies, they were indirectly representing foreign policy goals of their governments. BP, Statoil and Total with the strong political support from their governments were interested in becoming energy majors in the whole Eurasia and also were involved in geopolitical and strategic game targeting the region-building process in the Caspian Basin (Rzayeva, 2012). In fact, political interests of the home countries to minimize strong Russian influence and prevent Iran participation in the energy politics were prevailing over commercial preferences of the companies during the first phase of the Caspian energy development.

Involvement of the energy firms within the energy politics differs from oil sector to gas sector. As gas has been used as political leverage the engagement of the energy firms also are more politicized. In the case of SD field development foreign state companies such as Turkish TPAO, Russian Lukoil and Iranian NICO with relatively small shares act as national representatives to monitor and protect national interests of their governments through the whole decision-making process. Each of them pursues different political objectives. For Ankara

---

58 Initial shareholders in SD consortium were BP (25.5%), Statoil Hydro (25.5%), Total (10%), LukAgip (10%), SOCAR (10%), NICO (10%) and TPAO (9%).

59 In 2015 Statoil completed sale of its share in Shah Deniz consortium to Malaysian oil and gas company Petronas. Following this transaction the State Oil Company of Azerbaijan, SOCAR, will assume commercial operatorship. See: Rostad, K. (2015, April 30). *Statoil completes sale of 15.5% share in Shah Deniz to PETRONAS.* Retrieved January 4, 2016, from Statoil: http://www.statoil.com/en/About/Worldwide/Azerbaijan/Pages/ShahDeniz.aspx.

it is important to become an influential energy power bridging Western energy markets with the Caspian energy reserves as a part of its general energy strategy. For Russia and Iran it is critical to maintain their influence in the region's energy constellation.

Many things have changed since PSA agreement was signed among Azerbaijan and energy firms. Some of old partners left and some new energy firms entered into the SD consortium. Besides, previous consortium members were able to increase their shares. In 2013 Statoil sold 10% of its stake in SD consortium to BP and SOCAR, which got 3.3% and 6.7% respectively. In 2014 Total SA sold its 10% share to Turkish TPAO. In 2015 Statoil completed sale of its share in Shah Deniz consortium to Malaysian oil and gas company Petronas. Now BP with 28.8% of shares operates the consortium. Other shareholders include TPAO (19%), SOCAR (16.7%), Petronas (15.5%) Lukoil (10%) and NIOC (10%.)

The recent changes those took place within the consortium and intensification of the participation of the state energy companies in the decision making process had direct impact on sifting pipeline dynamics and energy politics. While analyzing decision-making process and reconfiguration of the shares within the Shah Deniz consortium, it becomes clearer that if some decisions have been motivated by political objectives, some have been driven from commercial rationale of leading energy companies involved in project development. In the light of the energy firms' power of influence during the decision-making process, it is possible to argue that interests of SOCAR and BP, which were looking for additional assets in the midstream and downstream projects, were crucial in setting rules of pipeline politics for the southern gas corridor.

6.3.1   Intertwined interests of SOCAR and BP

Successful realization of the oil export projects has had positive impact on development of Azerbaijan State Oil Company. During the second stage of the Caspian energy development SOCAR has emerged as an energy producing company with the new ambitions turning itself into international company with the huge technical and financial capacity. Today, in order to maximize its profits, the company is investing in various downstream projects and expanding its revenues through participating in the strategic energy projects in the region. SOCAR together with BP increased their shares within the Shah Deniz consortium, which increased their powers in shaping current energy politics in the region. Furthermore, growing financial and technical capacity of SOCAR was implying that the company would not be satisfied with a 10 percent stake in the SCP pipeline and would put all efforts to increase its shares along the supply chain (Rzayeva, 2012).

BP is the biggest energy firm, which has started operation in Azerbaijan's energy sector from the independence. Increasing its shares in SD consortium it has got advantage over the decision-making process regarding SD gas transportation. Notwithstanding BP is western energy company, it is had to say that it was representing interests of the EU during the second phase of the Caspian energy development. From the beginning of pipeline politics within the southern gas corridor, it has had different interests and pursued chiefly commercial ones, rather than political.

BP and SOCAR interested in increasing its revenues from the transportation of natural gas, which is considered a strategic commodity, through the southern corridor. They were looking for additional assets in the midstream and downstream projects. The control of the transportation infrastructure and transmission network becomes an urgent issue for the both energy companies, since it would enable companies to acquire more assets in the midstream project under advantageous terms. This explains why BP has proposed SEEP project in 2011 and SOCAR was looking for a getting share in the other pipeline projects proposed within the southern corridor. Interests of BP and SOCAR were intertwined.

Despite the resemblance of the interests, there were some disagreements between the two partners. Each had different approach regarding SD gas transportation route and proposed pipeline projects, SCP upgrade and the capacity of TANAP. BP has been mostly interested in small scale, low fixed infrastructure with minimum investment and maximum short-term returns. Azerbaijan, on the other hand, is interested in scalable projects, considering the gas reserves in the fields that are currently under development and beyond. Azerbaijani officials pointed out that new gas transportation infrastructure should be constructed or upgraded in away that to able to "service all these fields and be upgraded as the necessity arises" (Pirani, 2012).

Both companies have had different approach to the upgrading SCP line and TANAP's export capacity. In order to maximize its revenues from the new energy projects and expand its shares, SOCAR was advocating expansion of the SCP line, which would be able to transport additional gas from Azeri fields and Central Asia. Expansion of the infrastructure will allow company to increase its assets. Given that Azerbaijan would be able to control and operate the huge part of the strategic project. At the beginning BP was suggesting a 42-inch pipeline with 16 bcm capacity, which can be expanded to 22 bcm maximum. However, for Azerbaijani side it was not the best solution, considering the expected grow of the local natural gas production and potential export from the Central Asia. In contrast, SOCAR was suggesting a 56-inch pipeline with maximum 60 bcm capacity, which would be wholly consistent with the TANAP. After the long negotiation partners agreed to expand existing 7 bcm/ year SCP system in a way to accommodate a further 16 bcm per year with a new 48-inch pipeline loop, constructed parallel to the existing SCP line[60] (BP, 2016b).

Initially, SOCAR officials were planning TANAP with 31 bcm/year capacity, with scalability up to 60 bcm/year. The implication of TANAP having a capacity of 31 bcm/year is that, in addition to the Shah Deniz II gas, another 15 bcm/year will be available for export (Pirani, 2012). Furthermore, since Azerbaijan has been planning to become an energy transit country for Turkmen gas and also was engaged in negotiation with Ashgabat, it was obvious that SOCAR as a representative of the state's interest would encourage construction of the bigger pipeline. For Baku TANAP has been considered as means for realization commercial and political interests in the region.

Conversely, BP has had totally different vision. For BP construction of the TANAP with start-up capacity of 31 bcm per year and scalable up to 61 bcm per year would mean having similar challenges like Nabucco-classic, which BP did not supported from the beginning. Even if Azerbaijani total natural gas production would increase up to 50-55 bcm/year by 2025, there will be no sufficient volumes of gas for export, as domestic consumption, the amount of re-injected[61] and flared gas will increase. In fact it was not believed that incremental Azeri gas production could exceed Socar's expectations by such a huge margin by the projected start-up date (Pirani, 2012).

---

60   The SCPX project also includes construction of a number of facilities. These comprise two new compressor stations in Georgia, two intermediate pigging stations (one each in Azerbaijan and Georgia), six 48-inch block valve stations (one in Georgia and five in Azerbaijan), pressure reduction and metering stations at the international borders, and the interconnection with TANAP at the Georgia-Turkey border.

61   According to Simon Pirani, it is likely that extra reinjection will be required to maintain pressure in the ACG field in a decade. See: Pirani, S. (2012). *Central Asian and Caspian Gas Production and the Constraints on Export*. Oxford: OIES.

It is planned that TANAP project will start operation with the first gas flow from SD II in 2018. For that time only gas from SD II will be available for the exporting through the upgraded SCP line and TANAP. Considering the available volumes of the natural gas, in 2015 partners agreed to construct TANAP with the initial capacity 16 billion cubic meters per year which will be gradually increased to 24 billion cubic meters and then 31 billion cubic meters (Kok & Dag, 2015).

6.4 Route selection process

The determination of the transportation route was one of the significant decisions within the SD consortium, which took several stages and eliminations. It was expected that partners would make a final decision and choose the export route by the mid-2012. Four projects, Nabucco-West, TAP, ITGI and SEEP, were ready to provide a continuation route for TANAP and competed for the right to bring gas to Europe in a *winner-take-all contest* for SD II gas. Three distinct and partially conflicting interests of producer companies, consumer countries and transit and trading partners were presented within the framework of these projects. In order to ensure a fair and transparent selection process eight principles were set as the initial criteria by SD consortium members (BP Caspian, 2011):

- *Commerciality* – based principally on full export chain value, including market prices and infrastructure access charges and tariffs;
- *Project deliverability* – technical and organizational capability to execute the project plans on schedule and within budget;
- *Financial deliverability* – ability to cover development costs through equity, loans, grants or other funding;
- *Engineering design* – scope and quality of the engineering plans;
- *Alignment and transparency* – willingness to cooperate technically with Shah Deniz and to align with the timeline of Shah Deniz FFD;
- *Operability* – the long-term capability to manage physical and commercial operations safely, efficiently and reliably;
- *Scalability* – the potential for expansion or addition of export facilities to allow transportation of increased volumes as further gas supplies become available;

- *Public policy considerations* – meeting the EC's stated objective of enhancing supply diversity of European natural gas markets, and ensuring sustained support from all stakeholders.

In addition to these principles, partners agreed that any export route would need to have the ability to meet all relevant environmental, safety, social, legal and regulatory standards (BP Caspian, 2011). Selection of the pipeline project was also influenced by the natural gas markets along the transportation route.

In February of 2012 based on the selection criteria and referring to the financial situation of ITGI's two shareholders, Italian Edison and Greek state-owned DEPA the consortium's representatives announced that the ITGI was eliminated from the contest (Socor, 2012e). Given that, TAP took advantage over the route transporting Azerbaijani gas via the western Balkans to Italy. Three pipeline projects remained at the race continued the competition, whereas SEEP got strong support from BP and reconfigured Nabucco-West from the Commission. For the EU it was important to diversify natural gas supply in the countries over-dependent on Russian Gazprom. In this case regardless of TAP project was supported by the Commission, it was not considered as a priority. Different preferences of the partners were causing disagreements within the SD consortium.

6.4.1 Nabucco-West vs. SEEP

The EU backed Nabucco-West project was competing against BP's SEEP project. The transit routes of the both pipeline projects overlapped for the most part from Turkish border into Central Asia. Even though BP proposed SEEP was cost saving, the project was unscalable and did not responded to the energy interests of Azerbaijan. Following Azerbaijan-Turkey inter-governmental agreement on June 28, 2012 the Shah Deniz gas producers' consortium in Azerbaijan announced that it selected the Nabucco-West pipeline project to be the route for Caspian gas into Central Europe (Socor, 2012g). According to their June 28 communique, Shah Deniz producers selected Nabucco-West owing to its "greater maturity", compared with other transportation options (ibid.).

Selection of Nabucco-West over BP's SEEP was driven from Baku's and Ankara's decision to build the pipeline with scalable capacity for potentially large volumes from other sources, including Shah Deniz field and reserves from Central Asian producers. In comparison with Nabucco project, SEEP envisaged comparatively small volumes leaving almost no space for the additional sources.

It worth to underline that Azerbaijan and the EU supported scalable and strategic project based on commercial and political interests. In this case Nabucco became an optimal solution for TANAP's continuation into Europe. Moreover, reconfigured new pipeline was considered matured, cost-effective and bankable, which made it more attractive, at the same time justified its selection by the consortium, including BP.

6.4.2 Nabucco-West vs. TAP

After elimination of the two pipeline projects in 2012 TAP and Nabucco-West continued the competition over 10 bcm of natural gas from Shah Deniz field. Both were considered as matured projects by the consortium. With the entrance of the trans-Anatolian pipeline project into the pipeline game, Nabucco and TAP shareholders presented new versions of their projects as a continuation of the TANAP into the European markets.

Even TAP was a non-strategic project in terms of market orientation and volumes at the beginning, it was threatening Nabucco-West after reconfiguration. New designed TAP transmission capacity has become scalable up to 20 bcm. Being a corporate business project it has planned to build a longer overland pipeline across Greece and to link up with Italy's transmission pipelines operated by Snam Rete Gas, which transmission pipelines run the length of the Italian peninsula and connect with natural gas network in Switzerland. Besides using Italy as a gas hub TAP shareholders planned to reach other European countries through the Italian transmission system. Another advantage of the pipeline project was the possibility to reach gas markets in Western Balkan countries through Albania with construction of new interconnectors.

In contrast to TAP Nabucco-West was a strategic pipeline project aiming to deliver gas to the Central European Gas Hub (CEGH) in Baumgarten, Austria. It was also promising certain business opportunities for Shah Deniz shareholders, since CEGH is among the most important trading gas hubs in Europe. As distinct advantages, it inherited the inter-governmental agreements, project support agreements, and uniform European legal-regulatory regime, already established with the earlier configuration of the project (Socor, 2012d). Nabucco-West was regulated by the European Law and considered involvement of the other shareholders based on TPA.

As it was mentioned before, from the beginning the Nabucco project got strong political and financial support from the European Commission, despite the EU had changed its official position stating that it would welcome any route that would bring Caspian gas to Europe. For Caspian gas producers, namely for Azerbaijan, Turkmenistan and Kazakhstan it offered shortest route to lucrative European gas markets with an opportunity to earn higher netback prices from gas trade.

In fact, Nabucco-West was meeting the EU's strategic objective of supply diversification goals targeting vulnerable and dependent gas markets in Central and Eastern European countries, in comparison with TAP project. Moreover, these markets are not able to ensure availability of LNG, which increases comparability of pipeline gas and price. Nabucco-West was considered strategically important because of pricing, volumes and market orientation. Therefore, it was favored by the European Commission and of Caspian gas producing states.

It took almost a year after elimination of ITGI and SEEP, when the final decision was made. The decision-making process was very difficult and signaled different disagreements among the consortium members. Positions of the Shah Deniz shareholders and state actors were conflicting regarding the route preference. The biggest shareholders of the SD consortium, BP and Statoil, were advocating for selection of TAP project and State Oil Company of Azerbaijan was expressing interest in reaching the Baumgarten hub (Socor, 2013). For BP and Statoil exporting SD gas to Europe was a business opportunity, when for Azerbaijan it has represented an element of national interests and an investment into the country's future.

Despite existing political support for Nabucco-West, on June 26, 2013 the Shah Deniz consortium members declared selection of TAP project of the EU backed Nabucco-West pipeline project. The decision was influenced by certain factors. First, few days before the final decision, Azerbaijani SOCAR has acquired control of a gas network in Southeastern Europe by winning the tender for the Greek DESFA's (Public Gas Transmission System Operator) pipelines, beating Russian Gazprom and the smaller Russian Sintez (Assenova & Shiriyev, Azerbaijan and the New Energy Geopolitics of Southeastern Europe, 2015). Getting 66 percent of the Greek DESFA gave SOCAR strategic advantages and additional business opportunities in Greece and Balkan countries. Second, even after reconfiguration of old Nabucco pipeline project, Nabucco-

West was still facing lack of guaranteed supply sources and financing (Socor & Czekaj, 2015). Another challenging factor for Nabucco-West was withdrawal of German RWE from the Nabucco consortium by early 2013 and Bavaria's gas trader Bayerngas's cancelation of negations on future gas purchases from Nabucco. These influenced Nabucco negatively causing lost a guaranteed German customer and a possible connection with the Czech gas market from Baumgarten (Socor, 2013b). For Azerbaijan it was the end of market opportunities in northward and westward directions from Baumgarten and influenced SOCAR decision in favor of TAP. Contrary to Nabucco, with the entry of Swiss Axpo (replacing its Swiss EGL subsidiary), German Ruhrgas, and Belgian Fluxys[62] into the TAP consortium, the chances of the Azerbaijani gas to supply Switzerland, Germany, France and Great Britain with gas via Italy increased.

Finally, the decision was affected by the opportunity to be involved along the supply chain. In contrast to Nabucco, TAP consortium was offering shares to Shah Deniz producers. Following the final decision, on July 30, 2013 Azerbaijan's State Oil Company, BP, and Total of France entered the TAP pipeline consortium[63]. BP and SOCAR have each taken a 20% share while Total has acquired 10% (TAP AG, 2013). These developments meet mainly commercial interests of the energy companies and in some instance of Azerbaijani government, which is interested in getting commercial and strategic advantages from the realization of the southern gas corridor. In addition it was selection not only transportation route, but at the same time determination of the future markets for the Azerbaijani gas beyond the Shah Deniz.

6.5 Supply of Southeastern European countries

With the selection of the Trans-Adriatic Pipeline as a main export route for Azerbaijani gas increased the strategic significance of the Balkan states and added new dimension to the southern gas corridor. The natural gas supply from the Caspian region will ensure a diversity of gas supply and reduce the risk of

---

62   Fluxys holds ownership stakes in the Transitgas pipeline that links Switzerland with Germany, as well as in the Belgium-Britain and Netherlands-Britain interconnector pipelines.
63   Later Snam and Enagas joined the TAP and Statoil and Total left it. For today TAP's shareholding is comprised of BP (20%), SOCAR (20%), Snam (20%), Fluxys (19%), Enagás (16%) and Axpo (5%).

over-reliance on a single energy source in Southeastern Europe (SEE). TAP aims to supply energy markets in Greece, Albania and Italy. Moreover, the consortium plans to reach energy markets in Bulgaria, Bosnia and Herzegovina, Croatia, and Montenegro. In the future Slovenia, Hungary, Serbia and Macedonia can also benefit from TAP.

The new pipeline will have considerable impact on natural gas market and energy security situation in Southeastern Europe. In fact energy market in this part of Europe is small-scale, underdeveloped and not diversified. Countries will be supplied through TAP have limited fossil fuels and highly dependent on coal. The share of the coal in countries such as Serbia, Macedonia and Bosnia-Herzegovina constitute more than 50 percent of total consumption. Some other countries in the region, like Albania and Montenegro do not have natural gas within its energy mix due to the absence of required infrastructure (Akhundzade, 2015; TAP AG, 2016). As the EU is targeting to reduce carbon dioxide emissions and promote green energy, adding natural gas into the energy mix of Balkan countries becomes part of national energy strategy of these countries. Successful realization of the TAP project can make a positive change in the energy sector of the Balkan states and consequently meet Europe's energy security objectives.

The selection of TAP line as a main transportation route for the southern gas corridor has shifted priorities of the European Union. The purpose of the EU's Energy Community is to extend the internal energy market of the EU to Southeast Europe and beyond by pursuing and supporting the implementation of the relevant EU *acquis communautaire*, including the development of a complementary regulatory framework and the liberalization of the national energy markets in line with the acquis (Cutler, 2014). Countries that will be supplied via the new pipeline and interconnectors can be classified as underdeveloped and dependent on single supplier, namely from Russia.

At the initial phase, when TAP will become operational the direct beneficiaries from the line will be Greece and Albania. Greece is highly dependent from the Russian gas and it imports more than half its consumed gas from Gazprom paying 478 USD per thousand cubic meters (Socor & Czekaj, 2015). With the materialization of the southern gas corridor Greece will be able to diversify its energy supply sources and become less dependent on Russia.

Albania is underdeveloped in terms of the natural gas market. Electricity generation needs of the country are provided by oil and coal. Oil dominates the primary energy supply of Albania with a share of 57% (Berger, 2015). Natural gas infrastructure is not developed and country does not import natural gas. However, with the realization of the SGC the geopolitical and economic situation in Albania will drastically change. First, the pipeline will transit the country from east to west earning transit fees for the state budget and linking it with the international and regional gas transportation and distribution networks. Second, TAP shareholders also consider the gasification of the whole country, which will significantly improve country's energy security (Lani, 2015). Third, the construction of the Ionian Adriatic Pipeline (IAP) in the future that will connect with the energy network in Croatia increases Albania's strategic meaning on the energy map of Europe. Finally, the biggest contribution to the country's economy will be $1.12 billion of foreign direct investment (Lani, 2015).

It seemed that with the cancelation of the Nabucco pipeline project the countries of the Central Europe lost their chance of supply diversification. Two countries, Romania and Bulgaria were a vital part of the EU supported Nabucco project. Although, the Southern Gas Corridor will not pass through the Bulgaria and Romania, these two countries will be essential in linking TAP to the Central European gas markets through a south-north connection, which is quickly shaping up under the name the "vertical gas corridor" (Assenova, 2015). For years Bulgaria and Romania have been at the center of the main regional energy projects and took central line within the European energy security. Weakly diversified energy markets of these countries, especially Bulgarian market, have constituted an important element of the Moscow's energy politics pursued within the framework of the South Stream.

In contrast to Bulgaria, which is almost fully dependent on Russian oil and gas supplies, Romania is self-sufficient and relies on domestic production. The growth of the domestic production in Romania lead to abandoning Russian gas imports in April 2015 (Economica, 2015). For Romania realization of the southern gas corridor and Nabucco pipeline was the part of the energy strategy aligned chiefly with the European interests (Assenova, 2015). Within the new context, TAP, which replaced Nabucco project, constitutes a strategic importance for Bucharest.

The third country that will directly benefit from the construction of the TAP is Bulgaria. In terms of energy Bulgaria is extremely dependent on Russia and also pays one of the highest prices for gas in Europe (Socor & Czekaj, 2015). In 2014 Bulgaria signed an agreement with the Shah Deniz consortium for the delivery of 1 bcm of natural gas starting from 2019 (Badalova, Azerbaijan to open new stage in European energy security, 2014), which is almost half of the Bulgaria's total consumption. For Sofia, this is the chance to escape the tight grip of Russian energy companies, which has long exercised economic and political influence in the country (Assenova, 2015).

Along with Bulgarian natural gas market Macedonian energy market can also benefit from the natural gas supply through the Greece-Bulgaria Interconnector. Macedonia like Bulgaria is one of the most vulnerable countries to Russian gas supply interruptions. In fact, landlocked Macedonia is an important country in Balkans. Peace and stability in Macedonia are indicative of the wider region's chances to develop prosperous economies and strong energy sector (Assenova, 2015). As it has been mentioned above in the future TAP shareholders also plan to supply natural gas markets in Serbia, Kosovo and Bosina-Herzegovina (Janjic, 2015), which are also extremely dependent on natural gas supplies from Russia. By replacing South Stream with Turkish Stream Russia plans to strengthen its energy monopoly in Balkans and through the energy market power to influence politics in Europe. However, with the TAP supplying Greece and Balkan countries, it is possible to make a step forward energy diversification in Balkan region and contribute to the EU's efforts to improve energy security situation in South Eastern Europe.

Given the energy vulnerability of the most Southeastern European and Balkan countries and dependency on Russian gas supplies, the stakes inherent in the Southern Gas Corridor become very high. In the light of this fact, diversification of the natural gas supply in the most vulnerable countries of Europe compose very important element of the EU's energy security objectives. For the EU realization of the fourth corridor has always been a geopolitical as well as economic project.

6.6 Trans-Adriatic Pipeline and the future of Turkmen gas

The main question after the final decision of the Shah Deniz consortium was what the next for the Turkmen gas will be. Natural gas supply form Turkmenistan in westward has constituted an integral component of the EU's energy politics within the framework of the southern gas corridor and at the same time supported by U.S.A. The European Commission was actively involved in negotiating conditions regarding deliveries from Turkmenistan through proposed Trans-Caspian pipeline. In fact, the pipeline politics in the region has represented highly complicated puzzle with a different and competing interests of the all actors involved, which played simultaneously together and at the same time, against each other.

After the gas transit agreements signed between Azerbaijan and Turkey the situation turned to be more tense and it has set certain level of ambiguity, since the volumes of the new pipeline project have been reduced and considered chiefly transportation of Azerbaijani gas. Indirectly capacity limitations of the new pipeline system have started challenging access of Turkmen gas to the southern gas corridor. However, after the signing Izmir agreements the EU made it clear that whichever pipeline option is chosen, "there must be mechanisms to allow for new gas from Turkmenistan, when gas from that source is ready" (Badalova, 2011).

Turkmenistan and Azerbaijan have been showing interest in energy cooperation with the European Union. For Baku the engagement of Turkmenistan in regional energy projects has had strategic and commercial advantages. However, for the energy companies operating in Azerbaijan, Turkmen gas also has been considered as rivalry. Iran and Russia have openly criticized the EU initiative on construction of TCP line by providing counterarguments and referring to the geological conditions and unresolved status of the Caspian Sea. Kremlin several times declared that the trans-Caspian pipeline should not go ahead without the approval of all five coastal countries — Iran, Azerbaijan, Kazakhstan, Turkmenistan and Russia. In return Turkmenistan argued that the legal uncertainty has not prevented countries from entering into bilateral agreements on use of the Caspian. The tension around the Caspian pipeline projects has been growing rapidly by the Shah Deniz shareholders coming closer to the deadline for the route selection, on the one hand, and with Russia promoting and progressing with the South Stream project, on the other hand. While Azerbaijan tried to maneuver among various alternative pipeline projects and Turkmenistan chose a waiting position, Iran and Russia continued opposing the construction of the submarine pipeline.

When Nabucco was canceled Ashgabat did not reacted to the decision of Shah Deniz shareholders. In fact, there always existed other options for Turkmen gas, namely growing Chinese energy market and reaching European energy market through the territories of Iran and Turkey. Furthermore, China and Turkey are involved in energy cooperation with Turkmenistan.

A year later after Nabucco's cancelation, on November 7, 2014, Ankara and Ashgabat took a decision towards signing an agreement, which outlined a deal for Turkmenistan to supply its gas to TANAP (Socor & Czekaj, 2015).

None of the details of the framework agreement have been revealed, and the two sides have not commented on how gas from Turkmenistan is to reach pipelines in Turkey in the current absence of TCP line (Gurt, 2014). Indeed Turkmenistan has a great potential to supply energy market in Europe providing 30-40 bcm of natural gas. Ashgabat considers and works on possible and alternative export routes in western direction, which does not exclude the transportation through Iran. In this case, construction of the TCP line may still stay on a table as a project. Another factor affecting energy politics of Ashgabat is Russian energy politics pursued in the Central Asia. Current shifts happening in Gazprom's energy policy directions create a window of opportunity for Turkmenistan to start new energy cooperation with other states.

6.6.1   New dimension of the Russian gas politics in Central Asia

Control over the transportation of energy resources from Central Asia to Europe is a crucial determinant in Russia's energy and pipeline policies. Russia's pipeline policies have long been designed to ensure energy power via control of regional transportation infrastructure. This strategy blocks strategic pipeline projects seeking to bypass Russia's territory in the east-west direction from the Caspian Basin, and focuses on re-exporting natural gas from Central Asian producers and maintaining strategic grip over the natural gas deliveries to the European markets.

From the early 1990s, Russia, as a key importer of Central Asian gas, took control over the Central Asia-Center gas pipeline system. It bought and resold huge volumes of natural gas from Turkmenistan and Uzbekistan to the Europe, enjoying a monopoly in the European energy market and monopsony in Central Asian. However, starting from 2009 Gazprom drastically reduced natural gas supplies from Turkmenistan and Uzbekistan. In 2008 the company bought approximately 40 bcm of natural gas from Turkmenistan, and almost 15 bcm from Uzbekistan; by 2014 the total volumes of exported gas from these countries had decreased to 10 bcm and 4.5 bcm respectively.

Despite the significance of the region for Russia's energy security, Gazprom has continued cutting back on purchases. During the Investor Day held by Gazprom in Hong Kong in February 2015, the company Deputy Chairman Alexander Medvedev announced that Gazprom plans to reduce the volume of gas purchases from Uzbekistan and Turkmenistan down to 10 billion cubic meters in 2015. The company plans to reduce Turkmen gas purchases from 10 bcm to 4 bcm, and Uzbek gas purchases from 4.5 bcm to 1 bcm. This decision did not come as a surprise, as in October 2014, Gazprom's Marketing and Trading Director Pavel Oderov announced company's plan to continue reducing volumes of imported gas, as part of the company's revenue maximization policy through optimization of domestic production.

So the key questions at this point were: why is Gazprom continuing to cut back on purchases from Central Asia? Which factors have affected this decision? What are the implications of the decision for Central Asian producers? The key argument presented by Gazprom officials is based on increasing domestic production. But this reasoning is somewhat problematic, since it fails to address the evident impact of other, more important factors, including the escalation of the Ukrainian crisis and shifting market dynamics.

Energy decisions are frequently determined by political and economic dynamics. In the case of Russia and Central Asia, the historical trajectory of the decision reveals multiple influences. Based on different factors and conditional variables, the decision to reduce purchases is best analyzed in two phases: from 2009 till 2014 and from 2014 until the present. However, this categorization does not entail that the second phase emerged as a continuation of the first phase.

The first decision on reducing natural gas purchases from Turkmenistan dates back to the pipeline explosion in 2009, which resulted in the decline of imports and damaged energy relations between Turkmenistan and Russia. Moreover, the beginning of Russia's energy relationship with China and the construction of the Central Asia – China Gas pipeline system have opened up a new market; with its huge demand for Turkmen gas, this market has been accorded higher priority by Ashgabat.

In the case of Uzbekistan, the situation is a bit different. Uzbekistan is the third largest natural gas producer in Eurasia. However, the growing national consumption and aging energy infrastructure have slowed production and hindered the export of natural gas to Russia. The decline of production has weakened Uzbekistan's position as reliable and stable supplier for Russia.

On the other hand, because of the rise of natural gas prices in Central Asia, reselling Turkmen and Uzbek gas became less profitable for Russia. Gazprom could optimize revenues by exploiting its own fields, instead of being a porter of gas for Central Asian suppliers. In sum, it is possible to argue that during the first phase, the decline in supply was driven by the internal interests of Russia, Uzbekistan and Turkmenistan.

The more recent decision to reduce purchases should be reviewed from a different perspective, whereby external factors, especially the current Russian-Ukrainian crisis, play a more decisive role. For Russia, cutting off the natural gas supply to Ukraine and the EU sanctions affecting energy sector have negatively impacted the demand side. Gazprom's statistics show that demand for Russian gas in the European markets has declined almost to 9%. A comparison of the data from 2013 and 2014 demonstrates that the volumes of exported natural gas from Russia to European markets have been decreased by 15 bcm. In 2013 Gazprom exported 162 bcm of natural gas to Europe, compared to 147.2 bcm in 2014.

Of course, the fall in European demand for Russian gas is not exclusively the result of the crisis in Ukraine. The warm winter of 2014 and the availability of alternative gas supplies in the form of LNG also have influenced the situation. Without a doubt, the ongoing Ukrainian-Russian conflict, political decisions aimed at weakening Russia's political and economic power, and the intensification of the EU's energy diversification policy have challenged Russia's market position by increasing uncertainty around European demand for Russian gas in the near future.

It can be concluded that the latest decision to reduce purchases from Turkmenistan and Uzbekistan has been caused by the decline of the actual demand for Russian gas. By cutting back on purchases Russia can balance the difference between high production and low demand. This analysis demonstrates that these two phases do not follow on from one another in terms of causality, since the determinant factors are of different origins.

6.6.2   Further supply opportunities

Consequently, the next question is: "where will the 10 bcm Central Asian gas surplus go?" Gazprom's decision opens new market opportunities for Turkmenistan and Uzbekistan in the light of growing energy production in the region, and signals changes in the energy policy priorities of Central Asian producers. Russia is not the only player in the region engaged in energy projects with the regional producers. As part of their energy security strategies, regional producers are developing multi-vector gas export policies and are showing interest in cooperation with China, EU, Iran and Turkey.

Following the Gazprom decision, the State News Agency of Turkmenistan reported that Ashgabat would increase exports of natural gas to China through the Central Asia – China Gas Pipeline trans- mission system. Moreover, according to Ria Novosti, Uzbekistan plans to export an additional 10 bcm of natural gas to China in 2015. The decision to raise exported volumes to China was reached during the fall of 2014.

China's increasing role in both the global energy market and the Central Asian region has caused a shift in market dynamics. Starting from the middle of the last decade, China has actively pursued a targeted pipeline strategy, transforming itself into the main consumer of the region's natural gas resources.

By comparing volumes of natural gas exported to Russia and China, we can see that the drop in natural gas exports to Russia has coincided with increased natural gas imports by China via the Central Asia-China Gas Pipeline system. The pipeline system has three operational lines in parallel, each running for 1,830 kilometers through Turkmenistan, Uzbekistan and Kazakhstan, with an overall delivery capacity of 55 bcm. Moreover, in 2013, Uzbekistan and China began construction of the fourth line, with an annual transmission capacity of 30 bcm.

Further, EU member states have several times stressed their interest in energy cooperation with Central Asian suppliers, especially Turkmenistan. The EU is trying to get Turkmenistan involved in the Southern Gas Corridor, in order to diversify its supply sources. However, political, commercial and legal barriers have impeded involvement of Central Asian suppliers in SGC. Now, in the light of increasing Turkmen natural gas production, Gazprom's decision can be considered as a window of opportunity for the EU. Indeed, the success in this regard depends on how effectiveness and intensity of the political actions undertaken by the EU and partner states involved in SGC.

At the moment, Central Asian suppliers are more interested in gaining access to the Asian market. Energy cooperation with China is more attractive for Central Asian producers, because political issues are not interlinked with commercial interests.

Political and economic factors affected the decision to reduce purchases of natural gas from Turkmenistan and Uzbekistan during the different time frames. However, it is difficult to link the causality of these decisions. The February 2015 decision flows from the decline of European demand for Russian gas as a result of the Russian-Ukrainian crisis, followed by Russia's decision to cut off gas supplies to Ukraine. Russia needed Central Asian gas to meet the growing energy demand in the EU. Now, in the light of the demand decline and uncertainty of future demand, it makes more sense to reduce the surplus, in this case, natural gas imports from Turkmenistan and Uzbekistan.

In contrast, Central Asian suppliers needed Russia, because the Central Asia-Center gas pipeline system was only the means for natural gas transportation. The construction of the Central Asia – China Gas pipeline system has minimized Russia's strategic importance for Turkmenistan and Uzbekistan in terms of energy politics, and continues to open new market opportunities for these countries.

The weakening of Russia's economic presence in Central Asia opens up space for interactions between other regional actors. Indeed, the Asian market holds more appeal than the European market for Uzbekistan and Turkmenistan. Moreover, growing energy demand in China and the increase in market shares of Central Asian suppliers within the Asian market provides certain level of sustainability, due to the absence of political concerns.

The entrance of Central Asian producers into the European energy market introduces some complications. As long as political factors continue to impede cooperation between the EU and Central Asian countries, Turkmenistan's participation in the Southern Gas corridor is unlikely. The success of the EU in this regard depends on the political strategies of member states. Additionally, the Russian factor should not be forgotten. As long as Gazprom's revenues are mostly dependent on the European market, Russia will continue to block the construction of the new pipeline system in the western direction.

7 Conclusion

This study has sought to explore the new natural gas supply route development process and the dynamics of the pipeline politics pursued by various actors in the framework of the Southern Gas Corridor initiative. This concluding part summarizes the study in some detail and provides the findings against the assumptions made and hypotheses presented in part 1.2.

Due to climate change and the growing share of the natural gas in the energy mix, gas has become competitive and more attractive than other fossil fuels in the European energy markets as a source of the relatively clean energy. Following the Russian-Ukrainian gas crisis (2006 to 2009) the EU has set the policy direction towards diversification of the energy supply sources and decarbonisation strategy to meet its energy demand, increase energy efficiency, and provide contribution to active climate policy. Given that the realization of the southern gas corridor is conceived as a strategic project for the delivery of natural gas primarily from the Caspian and the Middle East to the European markets and important step towards decreasing the dependency of the Southeastern European on Russian gas.

Rather than describing the current energy politics in the Caspian region as a part the geopolitical competition or New Great Game as it has been done before, the aim of this study was to analyze the interactions between state and non-state actors, including interwoven political and economic interests of all actors, and to find out which factors affect the pipeline dynamics in the region. This constitutes the central question of the research.

In a broader sense, the analysis shed light on different aspects of the decision-making process shaped by the changing nature of the energy policy priorities of the smaller producer countries and the commercial interests of the major energy companies involved in natural gas production. The following conclusions can be formulated based on the comparison of the factors shaping the first and the second stages of the Caspian energy development, and analysis of the key actors and their energy security and foreign policy priorities. During the second stage of the Caspian energy development regional states and major energy companies have played a more decisive role during the decision-making

© Springer Fachmedien Wiesbaden GmbH, part of Springer Nature 2018
S. Amirova-Mammadova, *Pipeline Politics and Natural Gas Supply from Azerbaijan to Europe*, Energiepolitik und Klimaschutz. Energy Policy and Climate Protection, https://doi.org/10.1007/978-3-658-21006-9_7

process. Also, the transportation route selection process has been mostly affected by the commercial considerations of the companies rather than by political objectives of the state actors. The decision-making process was slowed down, because of the diversity of preferences and market interests of the partners and existed uncertainty.

In order to get a subtle understanding of the logics of the actors' moves and a clear picture of the current pipeline dynamics, the study suggests conceiving the energy politics as a part of a figuration, where all actors are linked with each other through the network of interdependencies with overlapping as well as colliding interests, perceptions, and visions. In addition to figuration concept, the study applied structural change and radical geopolitics concepts to identify how the non-state actors and geo-economics influence the decision-making process. Economic and political factors affecting states' and energy firms' policy priorities were also examined through the application of these theories.

One of the main arguments looked at in the study was that there are two different levels of pipeline competitions interlinked with each other. The first level of the pipeline competition is happening within the Southern Gas Corridor, among the strategic and commercial pipeline projects competing for the gas from Shah Deniz II. The second level shaped by a political struggle taken place between Russia and Europe under the umbrella of the energy security. Predominantly focusing on the first level, the study found that the link between two levels directly affects the pipeline dynamics in the southern gas corridor. Moscow's energy and foreign policy interests were among the key factors leading to the delays in the decision-making.

The determination of the key actors involved became an important element of the research, since it explains the motivations and the drivers stimulating the players' moves. Chapters 4, 5 and 6 focused on stakeholder analysis and their interests. The analysis elucidated the difference between energy politics pursued at both stages by the actors. Reviewing the actors involved at the second stage of the Caspian energy development no big difference was noticed from the first stage. The energy and pipeline politics in the Caspian region, in general, involved regional states like Azerbaijan, Turkmenistan, Kazakhstan, Russia and Iran and non-regional states like Turkey, the EU, the USA and China. These state actors took part in the formation of the energy politics during the realization of the oil pipeline projects. However, certain differences were found in the degree of their engagement and interests. Some played a more active role, some were rather passive or had only a very limited capacity to influence the direction of the pipeline politics.

During the first phase of the Caspian energy development Azerbaijan, Kazakhstan and Turkmenistan were lacking financial resources and were not able to exploit oil and gas reserves by their own capacities, mostly because of technological constrains. The dissolution of the Soviet Union damaged the political system and economy in these countries and the political elites were engaged in power struggles. The new governments in the new states saw involvement of the external players into the energy production as a source of improvement of the economic situation, as the means of regime maintenance and stabilization. The lack of the financial resources in these states ended up with a more active engagement of the foreign states in investment in the energy projects and pipeline politics in the region in 1990s and the 2000s.

The most active players in the Caspian region during the first stage of development were Russia, Iran, Turkey and the U.S. Each actor had pursued strategic objectives driven by certain political interests. A power vacuum emerged after the collapse of the Soviet Union transformed the Caspian region into an area of huge competition over the control of energy resources between traditional regional actors and external major powers.

The South Caucasus and the Central Asia are strategically important regions for Moscow's near abroad policy. Energy and pipeline politics pursued by Russia in these regions, during the first stage of the Caspian energy development was a part of its national energy security framework, which had a direct impact on national economy. For a long time, Central Asian energy producers had been dependent on the Russian transportation network, which was restricting sovereignty of the new post-Soviet states by maintaining Moscow's direct influence on economic and strategic decisions. Official Kremlin was controlling not only energy resources of the region, but also had a control over the pipeline routes, which connected resources with the international markets. For the Russian political elite control of the energy resources and the pipeline network was fulcrum for its power restoration and the key to realizing competitive advantages abroad. Therefore, attempts to ease the Russian grip over the transportation routes were considered as a threat by the Russian government.

Similar to Russia, Iran's interests in the region's energy politics were also linked with its energy strategy. Offering the shortest and cheapest route to global markets for oil and gas from the Caspian to the world markets, Tehran's commercial interests went beyond just offering its territory for transit of Caspian oil and gas. It was looking for transferring the country into a regional hub, in order to turn Iran into a gateway to Central Asia and Caucasus. However, the rivalry between the U.S. and Iran, left Tehran out of the energy politics and radically limited Iran's involvement in the exploration and development of the Caspian resources. Tehran became engaged in the energy production in Azerbaijan and got only 10% in Shah Deniz consortium through the National Iranian Oil Company.

USA and Turkey pursued a clearly aligned policy to support development of the oil and gas industry in the newly independent states of the Central Asia and the Caucasus and the realization of the Southern Gas Corridor. The USA interests in the Caspian region were based on multiple aspects. Foremost it tried to decrease the dependence of the Caspian states from the Russian transportation infrastructure. Therefore, Washington was promoting the construction of the new east-to-west oil and gas networks, in order to enable the oil and gas producing Caspian and Central Asian countries to export their resources without traversing the territory of the Russian Federation or Iran. To boost the state-building process in the newly independent states and to limit political, economic and cultural influence of Moscow and Tehran, the US government was enabling economic and political independence of the regional states.

Ankara was a representative of the Western interests in the Caspian region during the first phase of the Caspian energy development, and was involved in direct competition with Russia and Iran. Following the collapse of the Soviet Union, Turkey sought the ways of increasing its influence in the South Caucasus and Central Asia as part of its regional strategy policy and commercial interests. In supporting the construction of the new pipeline projects through its territory, Turkey pursued not only to become the bridge between the energy producing countries of the region and the world market, but also to increase its role as major regional player between Europe and the Caspian. Ankara's active involvement in the development of the region's oil and gas reserves and transportation projects was stimulated by commercial and political interests.

However, it was not possible to realize a huge trans-regional pipeline project during the first phase of development, since all major powers avoided involvement in direct competition with each other and were equally interdependent. On the other hand, regional small powers chose the politics of counterbalancing. None of the newly independent states were ready to act in a way, which would challenge the Russian foreign policy and could be considered as threat to Moscow's interests in the region. Also, Moscow had certain leverages to influence the political situation in the states tackling with the conflicts, ethnic tensions and instability.

The situation with natural gas export from the region was developing through a different trajectory. The landlocked geography of the Caspian region constrains supply options for natural gas and requires pipelines for its delivery. Pipeline transportation involves transit states, and for upstream producers it becomes important to get access to transportation facilities in the mid-and downstream countries. Gas supply is by its nature generally restricted to transport via pipeline and the supply routes are therefore rather inflexible. In addition, transit lines are extremely vulnerable to political manipulation and economic pressure, which ma siphon off any profitability in what is a zero-sum game between the pipeline owner and the transit country (Stevens, 1996). In this case, reliability of the transit corridor composes the main pre-condition not only for upstream producers, but also for midstream and downstream consumers. Hence, realization of a new transportation corridor involves not only many players but, at the same time, is very costly and challenging.

The situation with natural gas has been slowly changing, starting in the early 2000s. Gas became the key element determining a new direction of energy politics in the region. In contrast to oil, natural gas trade is built on political interests and frequently affected by the changes of inter-state relations. The second and current stage of the Caspian energy development (from 2006 to present) is characterized by active participation of Azerbaijan, Turkey, Russia, the EU and the numerous energy firms. Along with political factors and changing market dynamics, two major developments, namely the discovery of the Shah Deniz gas field and the beginning of the EU's diversification strategy following the Russian-Ukrainian gas crisis pushed towards reopening debates and brought back discussions about the realization of the southern gas corridor.

With the discovery of the Shah Deniz Field, the core driver behind the Southern Gas Corridor concept shifted from Turkmenistan to Azerbaijan. The discovery of the natural gas fields in Azerbaijan's off-shore territories provided a strong commercial driver to implement the original US-objective to develop a large-scale transportation solution to link Caspian gas to European markets without the Russian involvement. The intention to construct a new transportation route became a part of diversification strategy and engaged traditional partners, namely Russia and Europe, into energy competition.

As over the rears, natural gas supply security became a high priority issue for the EU, it started to play a more active role advocating the realization of the

Southern Gas Corridor, which will bring gas from the Middle East and the Caspian region to South and Eastern Europe. This initiative coincides principally with Europe's search for alternative gas sources. The construction of the new, long distance, large-scale cross-border pipeline system from the Caspian and the Middle East developed into the "project of European interest". However, instability in the Middle East (and North Africa) has limited the ways of finding reliable and secured sources of alternative supplies from that region. Therefore, supply of natural gas from the Caspian region, primarily from Azerbaijan and Turkmenistan, has been considered as a milestone to start the realization of the Southern Gas Corridor.

When reviewing the process within the new corridor it is possible two distinguish two different directions, determined by political and economic interests of the parties. Furthermore, if it is looked at the engagement level of the key actors, the process can also be divided into two phases. The first phase is determined chiefly by the EU's strong political support and energy firms' active engagement in the development of pipeline projects for the southern corridor. The second phase is characterized by the discovery of the new gas fields in the offshore sector of Azerbaijan and the crucial decision taken by Azerbaijan and Turkey to construct TANAP, which dramatically changed the dynamics of pipeline politics.

For the EU, realization of the Southern Gas Corridor has strategic importance, and accordingly, it supported the Nabucco pipeline project with huge capacity targeting long-term objectives. The project details is explicitly presented in Chapter 5. The key priority was given to the development of the relations and promoting genuine energy partnership with Azerbaijan, Turkmenistan, Iraq, Uzbekistan and Iran. This corridor is projected to become a multi-source and multi-vector network connecting various natural gas fields with the increasingly import-dependent countries of Europe. The strategic significance of such corridor for the EU is threefold. First, it will connect the European consumers with gas-rich regions through an alternative, scalable and secured supply route. Second, it may decrease the monopolistic position of Gazprom in the Eastern and Central European energy markets, indirectly weakening political and commercial influence of Russia in European countries. Third, the new corridor will enable the EU to meet the energy demand in the coming decades, which is important in the light of the depletion of the conventional reserves.

The EU considered Nabucco project and construction of the Trans-Caspian Pipeline strategically important and key steps towards achieving objectives of the southern corridor initiative. Both projects are regarded by Moscow as challenging Russia's energy strategy and as a threat to its interests in Central Asia and also in the European energy markets. The politics of the southern corridor became intense with the announcement of the creation a joint venture between Russian Gazprom and the Italian company ENI to build the South Stream pipeline system across the Black Sea. Moreover, some EU member states, like Bulgaria, Hungry, Greece, and Romania also supported the Russia's initiative of the new pipeline system. Announcement of the South Stream and openness of some European countries to this initiative questioned the reliability of the state partners and lead to delays in the decision-making process. The rationale behind Russia's decision to construct first the South Stream and replacing it with the Turkish stream had the aim to prevent the realization of the huge pipeline projects from the Caspian region in western direction. In fact, Russian involvement in the energy politics has been driven directly by political and commercial considerations. Moscow's decision to enter pipeline politics with the huge scale projects is part of the commercial realpolitik pursued in the Caspian region and European market.

Successful realization of the South Caucasus Pipeline and the discovery of the gas fields put Azerbaijan, along with Turkmenistan, at the center of the natural gas and pipeline politics. Both Baku and Ashgabat showed interest in the realization of the Trans-Caspian Pipeline transmission system and contributing to the Southern Gas Corridor. However, Turkmenistan's unchanged traditional approach to pipeline politics based on the motto "zero financial burden, hundred percent effectiveness" was causing various impediments for the Southern Gas Corridor. Furthermore, Turkmenistan's position not to undertake any financial burden for the realization of the Trans-Caspian Pipeline left it a bit out of the decision-making processes with a limited power to influence the energy and pipeline politics, instead was increasing Azerbaijan's role within the Southern Gas Corridor.

Comparing Azerbaijan's energy politics and decisions taken during the first and the second stages of the Caspian energy development, it became obvious that Azerbaijan is ready to pursue a more active and independent energy policy with certain strategic goals with revenues received from the oil export. Financial independence granted with the petrodollars enabled Baku to have direct influence in the decision-making process on the pipeline projects within the Southern Gas Corridor and to present its preferences as a pre-condition during the negotiations. Following interest-based multidimensional policy official Baku is targeting to reach wider energy markets as an energy producing and exporting country. For Azerbaijan securing access to the open, transparent and liberated markets, such as the European gas market, as well as expanding and developing new export routes has becomr a priority in its energy strategy

Turkey, which is an active player in the region since the 1990s, is involved in the natural gas supply politics not as a representative of the Western interests, but pursue its national energy strategy and interests. Ankara prioritizes its energy interests in two particular directions. First, it has ambitions to become an energy hub not only at the regional level, but also at the international level, transforming itself into a strategic bridge between Eastern oil and gas resources and Western markets. Second, it aims to secure gas for its own domestic market. Turkey's long-term energy strategy is shaped by a broad vision, taking into account the need to maintain a balance between its geography, foreign policy and national energy demands.

Although the various states are actively involved in energy politics and provided political support for the realization of the strategic pipeline projects, the energy firms have played a more decisive role in determination of the new export routes. Especially business interests of BP and SOCAR, which were looking for additional assets in the midstream and downstream projects, were crucial in setting rules of pipeline politics for the Southern Gas Corridor. Both, BP and SOCAR, are interested in increasing their revenues considering natural gas a strategic commodity. The control of the transportation infrastructure and transmission network became an urgent issue for both energy companies, since it enables companies to acquire more assets in the midstream projects under advantageous terms.

Three hypotheses concerning the current pipeline politics were presented in the preparation of the research (Chapter 1.2.1). The first hypothesis reviewed all actors, their competing interests, moves, and the level of the interdependency and explains the shifts within the pipeline dynamics based on balance of power. The second hypothesis focused on commercial aspects of the pipeline politics. The third hypothesis elucidated the relation between market dynamics and pipeline economics. The following reflection summarizes what has been argued in Chapters 3, 4, 5 and 6 and tries to state to what extend the hypotheses are correct.

The analysis confirmed that both political and business interests of the key actors and their decisions/moves mutually shape the pipeline dynamics. The decision-making process happens based on cost-benefit analysis, because the players in general are rational actors. Referring to the theoretical framework all actors of the current energy politics compose together a dynamic figuration, which shaped by the decisions and moves of the all actors. Figuration is viewed as a moving picture where actors and their actions are united and constitute complementary parts of the web of interdependencies. Consequently, their actions and moves are interlinked and require complex analysis. Change in the decision or move of one actor influences the move of another actor within that figuration.

Pipeline politics is a multi-dimensional and complex game, where the ability to control the actions of a relatively weak player provides an opportunity to a stronger player to control the dynamics and to change the figuration, when there prevails uneven level of interdependency among the actors. In contrast, when balance of power is equally distributed among players, players have fewer chances to control actions of their opponents, the course of game and the changing figuration. Consequently, by decrease of power differences the dynamic of the figuration changes independently from personal plans of the players and results in the interweaving of moves. This explains how the construction of TANAP project was taken and why it led to the transformation of the Nabucco project.

Although energy, especially natural gas supply, has been considered as an integral part of national security and foreign policy, market dynamics play a central role in ensuring, enhancing and attaining energy security. Additionally, the changing economic environment has affected market dynamics in a way that markets become more independent and self-regulated. It led to a decline of state power in existing state-market constellations and eased the state's influence. Therefore, both political and commercial considerations of the state and non-state actors must be considered as significant factors in the negotiation processes.

Natural gas supply from the Caspian region to Europe has become highly politicized. It would, however, be wrong to exclude the significance of economic factors shaping pipeline politics. It can be concluded that Russia's, Turkey's and Azerbaijan's motivations in the current energy politics were driven by both rationales. A strict distinction between political power and economic power is almost impossible. Since politics and economic are interlinked with each other within the structural configuration, the relationship between the two was analyzed by focusing on the effect of political authority (not only states) on markets and conversely, of markets on those authorities. In conclusion, not only political decisions affect market dynamics, market dynamics also affect political decisions. The analysis of the pipeline politics in the Southern Gas Corridor showed that political decisions and economic actions constitute opposite sides of the medallion mutually influencing each other.

The answer to the question why transformation of the pipeline projects happened is it happened because the interests of the actors changed. The causes of the shift in the pipeline dynamics can be affected by internal and external factors and changes in the actors' positions and priorities. While reviewing the causes of transformation of the Nabucco and TAP projects the following arguments seems to be significant. The Azerbaijani-Turkish initiative to construct the Tran-Anatolian Pipeline led to the transformation of the strategic and commercial pipeline projects. The Nabucco project, which was a strategic pipeline in order to meet commercial requirements of the producers was transformed and reformulated its initial design. On the other hand, the TAP line, which has been considered as a commercial project also had to make some changes in its structure. By targeting markets in Southeastern Europe and the Western Balkans the project has got political and strategic importance.

The analysis of the natural gas supply route selection process proved that pipeline dynamics were not affected only by the actors' interests. Financial dependence and ability to attract funds for the construction of the pipeline project also have played a vital role along with political factors within the Southern Gas Corridor.

The second hypothesis has also been confirmed. The examination of the decision to construct the TANAP line, which replaced the eastern part of the Nabucco project shows that financial independence of the producer and transit countries enable them to change the dynamics in order to achieve strategic and commercial objectives. The financial independence and ability to invest seems to be one of the keys for the economic power. With the globalization of the market economy, states compete for market shares and for the means to create wealth rather than for power over more territory. Before, states used to compete for power as a means to wealth, they now compete for wealth as a means to power.

Nabucco's failure was not only determined by the financial challenges, but also with the weakening of the EU's common energy policy. The failure of Brussels' common energy policy to come forward with the strategic pipeline

project designed to bring huge gas volumes to Europe bypassing Russia, created an opportunity for Azerbaijan and Turkey to enter the pipeline race with the
TANAP project. The TANAP project became an inevitable game changer,
emphasizing the differences between strategic and economic preferences of
Europe, on one side, and Turkey and Azerbaijan, on the other. Baku and Ankara have sought to take a greater share of the control of transport and marketing
arrangements along the supply chain. Baku needed a new export route for its
future gas production, whereas Ankara acquired additional infrastructure for the
alternative sources of fuel. Furthermore, with this new pipeline project Azerbaijan's State Oil Company SOCAR wouls maintain control over the transportation and can sell its gas directly to European consumers on the Turkish border,
based on European gas market price, which will ensure the flows of revenues
from the direct gas trade.

For Turkey, TANAP's advantages are twofold: First, Ankara can easily
reach and meet the growing natural gas demand in the western industrial cities
of Turkey. Second, the launch of the Azerbaijan-Turkey pipeline project constitutes a leverage, which Ankara can use in the negotiations with Moscow, to
soften the terms of gas supply contracts and price setting.

. The findings from our analysis imply that market dynamics and economy
of the pipeline are positively correlated, whilst the future of the pipeline project
is determined by these two factors. Indeed, throughout history the significance
of the pipeline has changed in terms of power and politics.

In energy politics, states act either as risk-averse or risk-prone. Consequently, states are not only concerned about maximizing the profit under various conditions, but also motivated to grip favorable perspectives and avoid
significant loss. The logic behind governments' choice to support and forge
transit commitments can be explained by two conditions. The first relates to the
salience of returns on investment. While, in the case of transit pipelines, the
focus is on receiving 'normal profits'.

The final decision on the supply route selection and choosing TAP over
the Nabucco-West primarily were influenced by economic factors: First, SO-
CAR has acquired control of a gas network in Southeastern Europe by winning
the tender for the Greek DESFA's pipelines, which provides Azerbaijan with
strategic advantages and additional business opportunities in Greece and the
Balkan countries. Second, even after reconfiguration of the old Nabucco pipeline project, Nabucco-West was still facing lack of guaranteed supply sources
and financing. Third, through the TAP line Azerbaijan has got the chance to

reach energy markets in Italy, Switzerland, Germany, France and Great Britain. For Azerbaijan, it was not only the selection of the transportation route, but at the same time, the determination of the future markets for its natural gas supply. Finally, energy firms involved at the upstream production were offered to join the consortium, in order to improve or even maximize their profits.

The Southern Gas Corridor is composed as a multiple pipeline network system with various exit and entry points. Such constellation ensures energy security of the consumer states and also provides a security of demand for the upstream producers. The ability to connect with different energy markets through a multiple pipeline network matches with the concerns of the private sector interested in maximizing netback values and returns on investment.

While this study has shed some light on the issues and factors that shape current pipeline dynamics, more research is needed to understand how and why energy firms succeed or fail to ensure political support for their commercial projects.

Bibliography

Abdelal, R. (2010). *The Profits of Power: Commercial Realpolitik in Europe and Eurasia.* Harvard Business School.

Abushov, K. (2010). Regionalism in the South Caucasus from theoretical perspective: is the South Caucasus a region? *Caucasus International , vol. 1* (2), 167-177.

Agnew, J., & Corbridge, S. (1989). The new geopolitics: the dynamics of geopolitical disorder. In P. Taylor, & R. Johnston (Ed.), *The world in "crisis": Geographical perspective* (pp. 266-288). Oxford: Basil Blackwell.

Akhundzade, E. (2015, January 27). *Energy Security in South East Europe: Role of the Southern Gas Corridor.* Retrieved December 6, 2015, from Hazar Enerji Enstitusu: http://www.hazar.org/blogdetail/blog/_energy_security_in_south_east_europe_role_of_the_southern_gas_corridor_1068.aspx

Akil, H. (2003, December 5). Turkey's Role in European Security as the Epicenter of Regional Energy Routes. *Turkish Policy Quarterly .*

Akiner, S. (2004). *The Caspian: politics, energy and security.* London, New York: RoutledgeCurzon.

Aliyev, I. (2012, June 5). *Speech by Ilham Aliyev at the opening in Baku of the 19th International Exhibition and Conference "Caspian Oil and Gas: Refining and Petrochemicals – 2012".* Retrieved June 24, 2012, from President of Azerbaijan: http://en.president.az/articles/5172

Aliyev, I. (2012). Statement by Ilham Aliyev, president of Republic of Azerbaijan, at the opening in Baku of the 19th International Exhibition and Conference "Caspian Oil and Gas: Refining and Petrochemicals – 2012". Baku: Official web page of president of Azerbaijan.

Amineh, M. P. (2003). Caspian Energy: A viable alternative to the Persian Gulf? *European Institute for Asian Studies , 03* (03), 1-20.

Amineh, M. P. (2003). *Globalisation, Geopolitics and Energy Security in Central Eurasia and the Caspian Region.* Clingendael International Energy Programme. The Hague: CIEP.

Amineh, M. P. (2004). Towards Rethinking Geopolitics. *Central Eurasian Studies Review , 3* (1), 7-8.

© Springer Fachmedien Wiesbaden GmbH, part of Springer Nature 2018
S. Amirova-Mammadova, *Pipeline Politics and Natural Gas Supply from Azerbaijan to Europe*, Energiepolitik und Klimaschutz. Energy Policy and Climate Protection, https://doi.org/10.1007/978-3-658-21006-9

APERC. (2007). *A quest for energy security in the 21st century: resources and constraints.* Tokyo: Asia Pacific Energy Research Centre.

Arrighi, G. (1994). *The Long Twentieth Century.* London: Verso.

Ashman, S., & Callinicos, A. (2006). Capital Accumulation and the State System: Assessing David Harvey's The New Imperialism. *Historical Materialism , 14* (4), 107-131.

Assenova, M. (2015). Pipeline Junction: Bulgaria, Romani and Macedonia and the Southern Gas Corridor. In M. Assenova, & Z. Shiriyev, *Azerbaijan and the New Energy Geopolitics of Southeastern Europe* (pp. 273-322). Washington D.C.: The Jamestown Foundation.

Assenova, M., & Shiriyev, Z. (2015). *Azerbaijan and the New Energy Geopolitics of Southeastern Europe.* Washington D.C.: The Jamestown Foundation.

Badalova, A. (2014, June 19). *Azerbaijan to open new stage in European energy security.* Retrieved July 2, 2014, from Azernews: http://www.azernews.az/analysis/68159.html

Badalova, A. (2011, October 27). *European Commission: Azerbaijani–Turkish deal to complement Nabucco intergov'l agreement.* Retrieved November 10, 2011, from Trend: http://en.trend.az/business/energy/1950254.html

Baev, P., & Øverland, I. (2010). The South Stream versus Nabucco pipeline race. *International Affairs , 86,* 1075-1090.

Bagirov, S. (1996, June). Azerbaijani oil: glimoses of a long history. Ankara, Turkey: Center for Strategic Reseach. Retrieved March 27, 2015

Bagirov, S. (2001). Azerbaijan's strategic choice in the Caspian region. In G. Chufrin, *The Security of the Caspian Sea Region* (pp. 178-194). New York: Oxford University Press.

Bahgat, G. (2003). *American Oil Diplomacy in the Persian Gulf and the Caspian Sea.* Gainesville, Fl., USA: University Press of Florida.

Bahgat, G. (2011). *Energy Security: an interdisciplinary approach.* Washington, DC: Wiley.

Bahgat, G. (2002). Pipeline Diplomacy: The geopolictis of the Caspian Sea region. *International Studies Perspectives* (3), 310-327.

Bahgat, G. (2007). Prospects for energy cooperation in the Caspian Sea. *Communist and Post-Communist Studies , 40,* 157-168.

Baker, J. A. (2000). *Running on Empty: Prospects for Future World Oil Supplies.* Institute for Public Policy. Houston: Rice University: BAKER INSTITUTE STUDY.

Bantekas, I. (2011). Bilateral Delimitation of the Caspian Sea and the Exclusion of Third Parties. *The International Journal of Marine and Coastal Law* (26), 47–58.

Baran, Z. (2002, Winter). The Caucasus: Ten Years after Independence. *Washington Quarterly*, *25* (1), pp. 221-234.

Baranick, M., & Salayeva, R. (2005, Jul 17). State-Building in a Transition Period: The Case of Azerbaijan. *ANALYSIS FOR NEW AND EMERGING SOCIETAL CONFLICTS*, pp. 208-220.

Berger, H. (2015). *Albania: Energy Efficiency*. Vienna: OeEB.

Bilgin, M. (2009). Geopolitics of European natural gas demand: Supplies from Russia, Caspian and the Middle East. *Energy Policy*, *37*, 4482-4492.

Bilgin, M. (2007). New prospects in the political economy of inner-Caspian hydrocarbons and western energy corridor through Turkey. *Energy Policy* (37), 6383–6394.

Biresselioglu, M. E. (2011). *European Energy Security: Turkey's future role and impact*. London: Palgrave Macmillan.

Bjoern, H., & Frederik, S. (2000). Theorising the Rise of Regionnes. *New Political Economy*, *vol. 5* (3), 457-472.

Blair, D. (2011, September 26). *Oil and Gas: BP plans gas pipeline to Europe from Azerbaijan*. Retrieved October 5, 2011, from Financial Times: http://www.ft.com/cms/s/0/ed9151b8-e84c-11e0-ab03-00144feab49a.html#axzz41G2cxAOf

Blanchard, O. (1997). *The Economics of Post-Communist Transition*. Oxford: Clarendon Press.

Block, F. (1987). *Revising State Theory*. Philadelphia: Temple University Press.

Block, F. (1994). The Roles of the State in the Economy. In N. Smelser, & R. Swedberg, *The Handbook of Economic Sociology*. Princeton: Princeton Univiverity Press/Russell Sage Foundation.

Bohi, D. R., & Toman, M. A. (1996). *The Economics of Energy Security*. Norwell: Kluwer Academic Publishers.

Bohr, A. (2004). Regionalism in Central Asia: new geopolitics and old regional order. *International Affairs*, *80* (3), 485-502.

BP Azerbaijan. (2016). *Shah Deniz*. Retrieved Jaunary 4, 2016, from BP in Azerbaijan: http://www.bp.com/en_az/caspian/operationsprojects/Shahdeniz/SDstage1.html

BP Azerbaijan. (2016). *South Caucasus Pipeline*. Retrieved February 12, 2016, from BP in Azerbaijan: http://www.bp.com/en_az/caspian/operationsprojects/pipelines/SCP.html

BP. (2015). *BP Caspian.* Retrieved May 20, 2015, from Baku-Tbilisi-Ceyhan pipeline:
    http://www.bp.com/en_az/caspian/operationsprojects/pipelines/BTC.html

BP Caspian. (2011, February). *Shah Deniz Consortium.* Retrieved August 25, 2011,
    from Principles for the selection of an export route to Eu rope for Shah Deniz gas:
    http://www.bp.com/liveassets/bp_internet/bp_caspian/bp_caspian_en/STAGING/l
    ocal_assets/downloads_pdfs/pq/Principles_for_selection_of_an_export_route_to_
    Europe_for_Shah_Deniz_gas.pdf

BP. (2016b). *South Caucasus Pipeline Project.* Retrieved March 1, 2016, from BP in
    Georgia:http://www.bp.com/en_ge/bp-georgia/about-bp/bp-in-georgia/southcaucas
    us-pipeline--scp-.html

BP. (2010). *Statistical Review of World Energy.* BP.

BP. (2014). *Statistical review of world energy 2014.* BP. London: BP.

Brenner, R. (1986). The Social Basis of Economic Development . In J. Roemer, *Analyti-
    cal Marxism* (pp. 23-53). Cambridge: Cambridge University Press.

Brill Olcott, M. (2006). International Gas Trade in Central Asia. In D. Victor, A. Jaffe,
    & M. Hayes, *Natural Gas and Geopolitics: From 1970 to 2040* (pp. 202-33).
    Cambridge: Cambridge University Press.

Brzezinski, Z. (1983). *Power and Principle: Memoirs of the National Security Adviser,
    1977- 1981.* New York: Farrar Straus & Giroux.

Brzezinski, Z. (1998). *The Grand Chessboard: American Primacy And Its Geostrategic
    Imperatives.* New York: Basic Books.

Burkitt, I. (1993). Overcoming Metaphysics: Elias and Foucault on Power and Freedom.
    Philosophy of the Social Sciences , 23 (1), 50–72.

Buzan, B., & Waever, O. (2003). *Regions and Powers: The Structure of International
    Security.* Cambridge: Cambridge University Press.

Cain, M., Ibrahimov, R., & Bilgin, F. (2012, October). Linking the Caspian to Europe:
    Repercussions of the Trans-Anatolian Pipeline. *Rethink Paper, 06.*

Callinicos, A. (2003). *The New Mandarins of American Power: The Bush
    Administration's Plans for the World.* Cambridge: Polity.

Campbell, C. (1997). *The coming oil crisis.* Essex: Multi-Science Publishing.

CEPS. (2011). *SECURITY OF ENERGY SUPPLY: A QUESTION FOR POLICY OR
    THE MARKETS?* Brussels: Centre for European Policy Studies.

Chazan, G. (2013, December 16). *Oil&Gas: Total and Statoil pull out of Tanap gas pipe
    deal.* Retrieved December 18, 2013, from Financial Times: http://www.ft.com
    /intl/cms/s/0/2d2e749a-666d-11e3-8675-00144feabdc0.html#axzz42V1T3DnT

Chester, L. (2009). Does the Polysemic Nature of Energy Security Make it a 'Wicked' Problem?

Chufrin, G. (2001). *The Security of the Caspian Sea Region.* Oxford: Oxford University Press.

Cohen, A. (2001). Is Russian New Caucasus Policy threat to Turkish Interests? *Eurasian Studies , 20* (Spesial Issue), 115.

Commission of the European Communities. (2006, March 8). A European Strategy for Sustainable, Competitive and Secure Energy. *Commission Green Paper .* Brussels. Retrieved June 23, 2012, from Energy:http://europa.eu/legislation_summaries/energy/european_energy_policy/l2 7062_en.htm

Commission of the European Communities. (2010). *Energy infrastructure priorities for 2020 and beyond – A Blueprint for an integrated European energy network.* Brussels: EC.

Commission of the European Communities. (2011). *On security of energy supply and international cooperation – "The EU Energy Policy: Engaging with Partners beyond Our Borders".* Brussels: EC.

Commission of the European Communities. (2008). *Commission Decision: establishin the projects of common interest eligible in the area of the trans-European energy networks selected for receiving Community financial aid in the framework of Desision C(2007)3945 for the annual work programme 2007 and the call for proposals launched on 15 June 2007.* Brussels: EU Commission.

Commission of the European Communities. (2008, November 13). Second Strategic Energy Review. *Communication from the Commission to the European Parliament, the Council, the European Economic and Social Committee and the Community of the Regions , COM(2008) 781 .* Brussels. Retrieved June 16, 2012

Commission of the European Union. (2007, October 02). *Summaries of the EU legislation.* Retrieved January 12, 2016, from Eur-Lex: http://eur-lex.europa.eu/legal-content/EN/TXT/?uri=URISERV:l27037

Cornell, S., Tsereteli, M., & Socor, V. (2005). Geostrategic Implications of the Baku-Tbilisi-Ceyhan Pipeline. In F. Starr, & S. Cornell, *The Baku-Tbilisi-Ceyhan Pipeline: Oil Window to the West* (p. 152). Washington DC; Uppsala: Central Asia Caucasus Institute; Silk Road Studies Program.

Correlje, A., & Van der Linde, C. (2006). Energy supply security and geopolitics: A European perspective. *Energy Policy , 34* (2006), 532-543.

Cutler, R. (2014). *The role of the Southern Gas Corridor in prospects for European energy security.* Caspian Report.

Dasseleer, P.-H. (2009). *Gazprom: L'idéalisme européen à l'épreuve du réalisme russe* . Paris: L'Harmattan.

Devin, G. (1995). Norbert Elias et l'analyse des relations internationales. *Revue française de science politique* , *45* (2), 305-327.

Devlin, B., & Heer, K. (2010). The Southern Corridor: Strategic Aspects for the EU. In K. Linke, & M. Viëtor, *Beyond Turkey: The EU's Energy Policy and the Southern Corridor* (pp. 5-9). Berlin: Friedrich-Ebert-Stiftung.

Deutsche Bank Research (2104) *Weltwirtschaftlicher Ausblick. 29. Juli 2014*

Ebel, R. (1997). *Energy Choices in the Near Abroad: The Haves and Have Notes Face the Future.* Washington, DC: Center for Strategic and International Studies.

EBRD. (2003). *Transition report 2003: Integration and regional cooperation.* London: European Bank for Reconstruction and Development.

Economica. (2015, March 21). Romania plans to abandon Russian gas imports in April 2015.

Edison. (2010). *ITGI Overview.* Retrieved March 28, 2011, from Edison: http://www.edison.it/en/company/gas-infrastructures/itgi.shtml

Edison. (2015, July 30). *ITGI Pipeline.* Retrieved Januray 14, 2016, from Edison Italy: http://www.edison.it/en/itgi-pipeline

Edison. (2011). *Opening the Southern Gas Corridor through ITGI pipeline.* Rome: Edison.

Edwards, M. (2003, March). The New Great Game and the New Great Gamers: disciples of Kipling and Mackingder. *Central Asian Survey* , 83-102.

Effimov, I. (2000). The oil and gas resource base of Caspian region. *Journal of Petroleum Science and Engineering* , *28* (2000), 157-159.

Egenhofer, C., & Labory, S. (1998). *The Development of Competition in European Gas Market.* Brussels: CEPS.

Egenhofer, C., & Legge, T. (2001). *Security of Energy Supply: A question for policy or marktes?* Brussels: CEPS.

Egenhofer, C., Gialoglu, K., Luciani, G., Boots, M., Scheepers, M., Costantini, V., et al. (2004). *Market-based options for security of energy supply. INDES Working Paper No. 1.* Burssels: CIEP.

Ehteshami, A. (2004). Geopolitics of hydrocabons in Central and Western Asia. In S. Akiner, *The Caspian: politics, energy and security.* London and New York: RoutledgeCurson.

EIA. (2015, January). *Azerbaijan.* Retrieved December 1, 2015, from US Energy Information Administration: http://www.eia.gov/beta/international/country.cfm?iso=AZE

EIA. (2015, January). *Kazakhstan: International energy data and analysis.* Retrieved December 1, 2015, from US Energy Information Administration: http://www.eia.gov/beta/international/analysis.cfm?iso=KAZ

EIA. (2015, July). *Turkmenistan.* Retrieved December 1, 2015, from US Energy Information Administration: http://www.eia.gov/beta/international/analysis.cfm?iso=TKM

Elias, N. (2000). *The Civilizing Process: Sociogenetic and Psychogenetic Investigation.* (E. Dunning, J. Goudsblom, M. S., Eds., & E. Jephcott, Trans.) Oxford: Blackwell.

Elias, N. (1978). *What is Sociology?* . London: Hutchinson.

Elkind, J. (2005). Economic Implications of the Baku-Tbilisi-Ceyhan Pipeline. In F. Starr, & S. Cornell, *The Baku-Tbilisi-Ceyhan Pipeline: Oil Window to the West* (pp. 39-60). Washington DC; Uppsala: Central Asia-Caucasus Institute: Silk Road Studies Program.

Ericson, E. (2009). Eurasian Natural Gas Pipelines: The Political Economy of Network Interdependence. *Eurasian Geography and Economics , 50* (1), 28-57.

ESMAP. (2003). *'Cross-Border Oil and Gas Pipelines: Problems and Prospoects.* Washington, DC: : UNDP/World Bank.

Estrada, A. (2006). European Energy Security: Towards the Creation of the Geo-energy Space. *34* (18), 3773–786.

Eurogas. (2008). *Natural Gas Demand and Supply, Long Term Outlook to 2030.* Eurogas.

European Commission. (2014). *COMMISSION STAFF WORKING DOCUMENT: In-depth study of European Energy Security.*

European Commission. (2000). *Green paper: Towards a European strategy for the security of energy supply.* Brussels: , Commission of the European Communities.

Feklyunina, V. (2008). The 'Great Diversification Game': Russia's Vision of the European Union's Energy Projects in the Shared Neighbourhood. *Journal of Contemporary European Research , 4* (2), 130-148.

Fettweis, C. J. (2009). No Blood for Oil: Why Resource Wars are Obsolete. In G. Luft, & A. Korin, *Energy Security Challenges for the 21st Century* (pp. 66-77). Santa Barbara: ABC-CLIO LLC.

Freifeld, D. (2009, Aug. 24). The Great Pipeline Opera: Inside the European pipeline fantasy that became a real-life gas war with Russia.

Frieden, A. (1994). International Investment and Colonial Control: A New Interpretation. *nternational Organization*, *48* (4), 559–93.

Gökay, B. (2001). *The politics of Caspian Oil*. New York: Palgrave.

Gültekin Punsmann, B. (2012, June). A Step Ahead Towards the Stage of Maturation in Azerbaijani-Turkish Relations: The Trans Anatolian Pipeline. *Evaluation Note* (36).

Gelb, B. (2005). *Caspian Oil and Gas: Production and Prospect.* Congressional Research Service, Resources, Science and Industry Division. Washington DC: CRS.

Gelpi, C. F., & Grieco, J. M. (2008). Democracy, Interdependence, and the Sources of the Liberal Peace. *Journal of Peace Research*, *45* (1), 17-36.

Goldman, M. I. (2008). *Petrostate: Putin, Power, and the New Russia*. New York: Oxford University Press.

Grau, W. L. (2001). *Hydrocarbons and a New Strategic Region: The Caspian Sea and Central Asia.* Fort Leavenworth, KS: Foreign Military Studies Office. Fort Leavenworth: Global Security.

Gurt, M. (2014, November 7). Turkmenistan inks deal with Turkey to supply gas to TANAP pipeline.

Gvalia, G., Siroky, D., Lebanidze, B., & Iashvili, Z. (2013). Thinking Outside the Bloc: Explaining the Foreign Policies of Small States. *Security Studies*, *22* (1), 98-131.

Haas, M. d. (2006). *Geostrategy in the South Caucasus; Power play and energy security of states and organisations.* The Hague: Netherlands Institute of International Relations Clingendael.

Haghighi, S. S. (2007). *Energy Security: The External Legal Relations of the European Union with Major Oil and Gas Supplying Countries.* Oxford, Portland and Oregon: Hart Publishing.

Harman, C. (1991). The state and capitalism today. *International Socialism*, *2* (51), 3-54.

Harrison, P., & Westall, S. (2011, February 17). *EU pushes strategic gas pipelines to merge.* Retrieved February 19, 2011, from Reuters: http://www.reuters.com/article/eu-energy-pipelines-idUSLDE71G0TK20110217

Harvey, D. (2003). *The New Imperialism.* Oxford: Oxford University Press.

Hasanli, I. (2010). *Azerbaijan's Borders: Formation of the borders of post-Soviet Azerbaijan.* Baku: Centre for National and International Studies .

Hasanov, A. (1998). *Azerbaijan's foreign policy: countries of Europe and USA, 1991–96*. Baku.

Helm, D. (2003). *Energy, the State and the Market: British Energy Policy Since 1979.* New York: Oxford University Press.

Helm, D. (2007). *The Russian dimension and Europe's external energy policy.* University of Oxford.

Herzig, E. (2001). Iran and Central Asia . In R. Allison, & L. Jonson, *Central Asian Security: The New International Context,* (pp. 171-198). Washington, D.C.: Brookings Institution.

Honore, A. (2006). *Future natural gas demand in Europe.* Oxford Institute for Energy Studies, Natural Gaas Research Program. Oxford: OIES.

Hopkrik, P. (1994). *The Great Game: The Struggle for Empire in Central Asia.* NY: Kodansha International.

Hulbert, M. (2012). *Azerbaijan: knock, knock, knocking on Europe's door.* Brussels: European Energy Review.

ICIS. (2010, September 16). *RWE offers to merge Nabucco and ITGI projects – report.* Retrieved November 12, 2010, from ICIS: http://www.icis.com/resources/news/2010/09/16/9394008/rwe-offers-to-merge-nabucco-and-itgi-projects-report/

Idan, A., & Shaffer, B. (2011). The Foreign Policies of Post-Soviet Landlocked States. *Post-Soviet Affairs, 27* (3), 241-268.

IEA. (2015). *History.* Retrieved November 23, 2015, from International Energy Agency: http://www.iea.org/aboutus/history/

IEA. (2010). *International Energy Outlook 2010.* Retrieved May 13, 2011, from http://www.eia.doe.gov/oiaf/ieo/nat_gas.html

IEA. (2010). *Natural Gas Market Outlook at International Energy Outlook .*

IEA. (2008). *Development of Competitive Gas Trading in Continental Europe: How to Achieve Workable Competition in European Gas Markets?* Paris: International Energy Agency/OECD.

IEA. (2008a). *Key World Energy Statistics 2008.* International Energy Agency. Paris: IEA.

IEA. (2008b). *Development of Competitive Gas Trading in Continental Europe: How to Achieve Workable Competition in European Gas Markets?* Paris: International Energy Agency/OECD.

IEA. (2000). *World Energy Outlook 2000 .* Paris: IEA-OECD.

IEA. (2002). *World Energy Outlook.* IEA. Paris: International Energy Agency.

IEA. (1998). *Caspian Oil and Gas.* Paris: Organization for Economic Cooperation and Development.

IEA. (1995). *IEA, the First 20 Years: Major policies and actions.* IEA.

IGI Poseidon. (2008). *The Project: Strategic value.* Retrieved May 11, 2011, from IGI Poseidon: http://www.igi-poseidon.com/english/strategicvalue.asp

Isachenko, D. (2012). *The Making of Informal States: Statebuilding in Northern Cyprus and Transdniestria.* London: Palgrave Macmillan.

Ismailzade, F. (2002, November 19). Democratization Trends in Azerbaijan: Half Empty or Half Full?". *Russia and Eurasia Review , 1* (12).

Ismayilov, E. (2011, September 21). *Oil and Gas: SOCAR – Azerbaijan's proven gas reserves exceed 2.5 TCB.* Retrieved October 10, 2011, from Trend: http://en.trend.az/business/energy/1934571.html

Jaffe, A., & Manning, R. (1999). The Myth of the Caspian Great Game: the real geopolitics of energy. *Survival , 40* (4), 112-129.

Janjic, D. (2015). Serbia and Bosnia-Herzegovina: Prospective Markets for Caspian Gas. In M. Assenova, & Z. Shiriyev, *Azerbaijan and the New Energy Geopolitics of Southeastern Europe* (pp. 323-356). Wahsington D.C.: The Jamestown Foundation.

Jentleson, B. W. (1986). *Pipeline Politics: The Complex Political Economy of East-West Energy Trade.* Ithaca: Cornell University Press.

Jonson, L. (2001). The new geopolitical situation in the Caspian region. In G. Chufrin, *The Security of the Caspian Sea Region* (pp. 11-32). Oxford: Oxford University Press.

Jonson, L. (2002). The new geopolitical situation in the Caspian region. In G. Chufrin, *Security of the Caspian Sea Region* (pp. 11-32). Stockholm: SIPRI.

Jordan, H. (1982). *Crisis: The last year of the Carter presidency.* Putnam Adult.

Joskow, P. (2010). *Market Imperfections versus Regulatory Imperfections.* CESifo DICE.

Joskow, P. (2007). Supply security in competitive electricity and natural gas markets. In C. Robinson, *Utility Regulation in Competitive Markets: Problems and Progress* (pp. 62-77). Edward Elgar Publishing.

Kleveman, L. (2003). *The New Great Game: blood and oil in Central Asia.* New York and London: tlantic Books.

Kok, E., & Dag, D. (2015, March 13). *Major shareholder in the project of century.* Retrieved May 28, 2015, from TANAP: http://www.tanap.com/media/press-releases/major-shareholder-in-the-project-of-the-century/

Kruyt, B., Vuuren, D., Vries, H. d., & Groenenberg, H. (2009). Indicators for energy security. *Energy Policy , 2166-2181.

Kryukov, A., & Moe, A. (2007). Russian Oil Industry: Risk Aversion in a Risk-prone Environment. *Eurasian Geography and Economics* , 341–57.

Kusznir, J. (2013, February 18). TAP, Nabucco West, and South Stream: The Pipeline Dilemma in the Caspian Sea Basin and Its Consequences for the Development of the Southern Gas Corridor. *CAUCASUS ANALYTICAL DIGEST* (47).

Kydd, A. H. (2005). *Trust and Mistrust in International Relations.* Princeton: Princeton University Press.

Lani, R. (2015). Albania and Azerbaijan: New partnership, new possibilities. In M. Assenova, & Z. Shiriyev, *Azerbaijan and the New Energy Geopolitics of Southeastern Europe* (pp. 233-271). Washington D.C.: The Jamestown Foundation.

Locatelli, C. (2008). EU Gas Liberalization as a Driver of Gazprom's Strategies? *Russie.Nei.Visions* (26).

Locatelli, C. (2010). Russian and Caspian Hydrocarbons: Energy Supply Stakes for the European Union. *Europe-Asia Studies* , *62* (6), 959–971.

Luciani, G. (2004). *Security of supply for natural gas markets: what is it and what is it not.* INDES.

Lussac, S. (2010). A Deal at Last: A bright future for Azerbaijani gas in Europe? *Central Asia and Caucasus Institut Analyst* .

Luttwak, E. (1993). *The endangered American dream* . New York: Simon and Schuster.

Mairet, F. (2006). *New Stakes in the Caucasus and Central ASia.* Bloomington: AuthorHouse.

Mandil, C. (2008). *Sécurité énergétique et Union européenne : propositions pour la présidence française.* Paris.

Mankoff, J. (2009). *Eurasian Energy Security.* New York: Council on Foreign Relations.

Manning, R. (2000). The Myth of the Caspian Great Game and the new Persian Gulf. *The Brown Journal of World Affairs* , *VII* (2), 15-33.

Mansfield, E. R., & Pollins, B. M. (2003). *Economic Interdependence and International Conflict: New Perspectives on an Enduring Debate.* Ann Arbor:: University of Michigan.

Marácz, L. (2011). The strategic relevance of AGRI in Europe's Southern Gas Corridor. *Karadeniz Araştırmalari* , *7* (28), 19-28.

Martin, W., & Harrje, E. (2005). The International Energy Agency. In J. Kalicki., & D. Goldwyn, *Energy and Security: Toward a New Foreign Policy Strategy* (pp. 97-116). Washington: Woodrow Wilson Press.

May, C. (1996). Strange fruit: Susan Strange's theory of structural power in the international political economy. *10* (2), 167-189.

McCarthy, J. (2000). The Geopolitics of Caspian Oil. *Jane's Intelligence Review* .

McLellan, B. (1992). Transporting oil and gas – the background to the economies. *Oil and Gas Finance and Accounting* , *7* (2).

Mehdiyoun, K. (2000). Ownership of Oil and Gas Resources in the Caspian Sea. *The American Journal of International Law* , *94* (1), 179-189.

Meiertöns, H. (2010). *The Doctrines of US Security Policy: An Evaluation under International Law.* Cambridge, UK: Cambridge University Press.

Meister, S., & Viëtor, M. (2011). Southern Gas Corridor and South Caucasus. *South Caucasus 20 Years of Independence.* Tbilisi, Kaukasien <Süd>: Friedrich-Ebert-Stiftung.

Mercer, J. (2010). Emotional Beliefs. *International Organization* , *64* (1), 1-31.

Mercille, J. (2008). The radical geopolitics of US foreign policy: Geopolitical and geoeconomic logics of power. *Political Geography* , *27* (5), 570-586.

Mercille, J., & Alun, J. (2009). Practicing Radical Geopolitics: Logics of Power and the Iranian Nuclear "Crisis". *Annals of the Association of American Geographers* , *99* (5), 856 – 862.

Meyer, J., Boli, J., Thomas, J., & Ramirez, F. (1997). World Society and the Nation-State. *American Journal of Sociology* (103), 144-181.

Meyer, K., & Brysac, S. (2001). *Tournament of Shadows: The Great Game and the Race for Empire in Asia.* London: Abacus Books.

Miliband, R. (1983). State power and class interests . *New Left Review I* , *138*, 57–68.

MITEI. (2011). *The Future of Natural Gas – An Interdisciplinary MIT Study.* MIT Energy Initiative.

Morady, F. (2011). Iran ambitious for regional supremacy: the great powers, geopolitics and energy resources. *Journal of the Indian Ocean Region* , *7* (1), 75-94.

Nabucco. (2011). *Construction Principles.* Retrieved May 10, 2011, from Nabucco Gas Pipeline: http://www.nabucco-pipeline.com/portal/page/portal/en/pipeline/construction

Nabucco. (2010, January). *Pipeline Overview.* Retrieved October 12, 2010, from Nabucco Pipeline: http://www.nabucco-pipeline.com/portal/page/portal/en/pipeline/overview

Namazi, S., & Farzin, F. (2004). Division of the Caspian Sea: Iranna policies and concerns. In S. Akiner, *The Caspian: politics, energy and security* (pp. 208-220). London, New York: RoutledgeCurzon.

Nanay, J. (1998). The US in the Caspian: The Divergence of Political and Commercial Interests. *Middle East Policy*, *6*, 150-157.

NEA. (2012). *NEA Publications and Reports//The Security of Energy Supply and the Contribution of Nuclear Energy*. Retrieved February 21, 2013, from Nuclear Energy Agency: http://www.oecd-nea.org/pub/security-energy-exec-summary.pdf

Noël, P. (2013). *EU Gas Supply Security: Unfinished Business.* University of Cambridge, EPRG. University of Cambridge.

Nye, J. (2008). Public Diplomacy and Soft Power. *ANNALS of the American Academy of Political and Social Science*, *616* (1), 94-109.

O'Hara, S. (2004). Great game or Grubby game? the struggle for control of the Caspian. *Geopolitics*, *9* (1), 138-160.

O'Hara, S., & Heffernan, M. (2006). From geo-strategy to geo-economics: the 'heartland' and British imperialism before and after Mackinder . *Geopolitics*, *11* (1), 54–73.

Olcott, M. (2002). *Kazakhstan: Unfulfilled Promise.* Washington DC: Carnegie Endowment for International Peace.

Omonbude, E. J. (2007). The Transit Oil and Gas Pipeline and the Role of Bargaining: A Non-Technical Discussion. *Energy Policy, 35:* , 6188–94.

Ostrowski, W. (2010). *Politics and Oil in Kazakhstan.* Abingdon, Oxon: Routledge.

Oxford Analytica. (2011). *Azerbaijan seeks gas export route bypassing Turkey.* Oxford: Global Strategic Analysis.

Paul, A., & Grgic, B. (2010, October 27). Entering the end game: the race for Caspian gas. *EPC Commentary* .

Penrose, E., Joffe, G., & Stevens, P. (1992). Nationalization of foreign owned property for a public purpose: an economic perspective on appropriate compensation. *Modern Law Review*, *55* (3), 351-67.

Pflüger, F. (2012, January 12). *The Southern Gas Corridor: Reaching the Home Stretch.* Retrieved January 13, 2012, from European Energy Review: http://www.europeanenergyreview.eu/site/pagina.php?id=3455&print=1

Pirani, S. (2012). *Central Asian and Caspian Gas Production and the Constraints on Export.* Oxford: OIES.

Pirani, S., Stern, J., & Yafimava, K. (2009). *The Russo-Ukrainian gas dispute of January 2009: a comprehensive assessment.* OIES.

Polo, M., & Scarpa, C. (2003). *The Liberalization of Energy Markets in Europe and Italy.* Milano: IGIER – University Bocconi.

Pomfret, R. (2005). Kazakhstan's Economy since Independence: Does the Oil Boom offer a Second Chance for Sustainable Development? *Europe-Asia Studies , 57* (6), 859-876.

Pomfret, R. (2006). *The Central Asian Economies Since Independence.* New Jercy: Princeton University Press.

Pritchin, S. (2011). Azerbaijan's new gas strategy. *Turkish Policy Quarterly , 9* (2).

Rostad, K. (2015, April 30). *Statoil completes sale of 15.5% share in Shah Deniz to PETRONAS .* Retrieved January 4, 2016, from Statoil: http://www.statoil.com/en/About/Worldwide/Azerbaijan/Pages/ShahDeniz.aspx

Russett, B., & Oneal, J. (2001). *Triangulating Peace: Democracy, Interdependence, and International Organizations.* New York: Norton.

Russian Federation Decree No. 1715. (2010). *Energy Strategy of Russia: For the period up to 2030.* Moscow: Institute for Energy Studies.

Rzayeva, & Tsakiris. (2012b, November 8). BP, Total, Statoil to take stakes in TANAP . *Strategic Imperative .*

Rzayeva, G. (2012). A Complicated Corridor: Gas to Europe – it's not just economics. *Caucasus International, 2* (2), 141-159.

Rzayeva, G. (2010). Azerbaijan Eurasia's Energy Nexus. *Turkish Policy Quarterly , 9* (2), pp. 55-68.

Rzayeva, G. (2015). Azerbaijan-Turkey Relations and Baku's Energy Strategy for Southeastern Europe. In Z. Shiriyev, & M. Assenova, *Azerbaijan and the New Energy Geopolitics of Southeastern Europe* (pp. 170-202). Washington DC: The Jamestown Foundation.

Rzayeva, G. (2014, February). Natural Gas in the Turkish Domestic Energy Market: Policies and Challenges.

Rzayeva, G., & Tsakiris, T. (2012). Strategic Imperative: Azerbaijani Gas Strategy and the EU's Southern Corridor. *SAM Review.*

Saivetz, C. (2009). Tangled Pipelines: Turkey's Role in Energy Export Plans. *Turkish Studies , 10* (1), 95-108.

Schatz, E. (2013). *Modern clan politics: the power of blood in Kazakhstan and Beyond.* Seattle, WA: University of Washington Press.

Shaffer, B. (2010). Caspian energy phase II: Beyond 2005. *Energy Policy, 38,* 7209–7215.

Shaffer, B. (2009). *Energy Politics.* Philadelphia: University of Pennsylvania Press.

Shirinov, R. (2011). A Pragmatic Area of Cooperation: Azerbaijan and the EU. *International Politics and Society , 3* (74).

Shiriyev, Z. (2015, March 11). Turkmenistan, Turkey and Azerbaijan: A Trilateral Energy Strategy?. *Eurasia Daily Monitor, 12* (45).

Socor, V. (2011c, November 1). Azerbaijan And Its Gas Consortium Partners Sign Agreements With Turkey . *EDM* .

Socor, V. (2012g, June 29). Nabucco-West Selected for Caspian Gas Delivery to Central Europe. *Eurasia Daily Monitor.*

Socor, V. (2012e, February 22). Shah Deniz Consortium Members Signal Conflicting Priorities (Part One). *Eurasia Daily Monitor, 9* (37).

Socor, V. (2013, June 27). Shah Deniz Gas Producers Select Trans-Adriatic Pipeline Route into Europe over Nabucco. *Eurasia Daily Monitor Volume , 10* (122).

Socor, V. (2011b, November 2). South-East Europe Pipeline: A Downsized Nabucco Proposed by BP. *Eurasia Daily Monitor, 8* (22).

Socor, V. (2013b, Hune 28). The Curtain Falls on Nabucco's Last Act. *Eurasia Daily Monitor, 10* (123).

Socor, V. (2012f, April 9). The Trans-Anatolia Gas Pipeline and Its Continuation Options to Europe. *Eurasia Daily Monito , 9* (70).

Socor, V. (2012c, Januray 6). Trans-Anatolia Gas Pipeline: Wider Implications of Azerbaijan's Project (Part Two) . *Eurasia Daily Monitor, 9* (4).

Socor, V. (2012a, January 5). *Trans-Anatolia Gas Pipeline: Wider Implications of Azerbaijan's Project.* Retrieved January 12, 2012, from Jamestown Foundation: http://www.jamestown.org/single/?no_cache=1&tx_ttnews%5Btt_news%5D=388 46&tx_ttnews%5BbackPid%5D=7&cHash=dc04cb9a31540c9f38bd052aac6cd360

Socor, V. (2012d, March 28). Trans-Anatolia, Nabucco-West Pipeline Projects: An Optimal Fit. *Eurasia Daily Monitor Volume, 9* (62).

Socor, V. (2011a, January 20). Turkmen President Supports Trans-Caspian Pipeline in Meeting With Top EU Officials. *Eurasia Daily Monitor Volume , 14.*

Socor, V. (2007, October 12). White Stream: Additional Outlet Proposed for Caspian Gas to Europe. *Eurasia Daily Monitor , 4* (189).

Socor, V., & Czekaj, M. (2015). Southeastern Europe's Energy Nexus: A Crossroads of Pipelines and Geopolitics. In M. Assenova, & Z. Shiriyev, *Azerbaijan and the New Energy Geopolitics of Southeastern Europe* (pp. 3-62). Washington D.C.: The Jamestown Foundation.

Startfor. (2011, February 22). *Azerbaijan's Position in Europe's Energy Diversification Plans.* Retrieved February 23, 2011, from Startfor Global Intelligence: http://www.stratfor.com/memberships/185466/analysis/20110221-azerbaijans-position-europes-energy-diversification-plans

Stern, J. (2005). *The Future of Russian Gas and Gazprom* . Oxford: Oxford Institute For Energy Studies.

Stern, J. (2007). *Is There a Rationale for the Continuing Link to Oil Product Prices in Continental European Long-Term Gas Contracts?*. Oxford Institute of Energy Studies. Oxford, UK: OSIE.

Stern, J. (2006). The Russian-Ukrainian gas crisis of January 2006. *OIES Working Paper NG* .

Stern, J. (2002). The Security of European Natural Gas Supplies.

Stern, J., Pirani, S., & Yafimava, K. (2015, January). Does the cancellation of South Stream signal a fundamental reorientation of Russian gas export policy? *Oxford Energy Comment* .

Stevens, P. (1996). A History of Transit Pipelines in the Middle East: Lessons for the future. *4th International Conference of the International Boundaries Research* (pp. 1-16). Dundee: Center for Petroleum and Mineral Law and Policy.

Stevens, P. (2009). *Transit Troubles: Pipelines as a source of conflict.* Royal Institute of International Affairs. London: Chatham House.

Stopford, J., Strange, S., & Henley, J. (1991). *Rival States, Rival Firms: Competition for World Market Shares.* Cambridge: Cambridge University Press.

Strange, S. (1995). The Defective State. *Daedalus , 124* (2), 55-74.

Strange, S. (1994). *States and Markets* (2nd Edition ed.). London: Bloomsbury Academic.

Strange, S. (1992). States, firms and diplomacy. *International Affairs, 68* (1), 1-15.

Strange, S. (1983). Structures, Values and Risk in the Study of the InternationalPolitical Economy. In R. Jones, *Perspectives on Political Econom.* London: Francis Pinter Publishers.

Stulberg, A. N. (2012, Jan 31). Strategic bargaining and pipeline politics: Confronting the credible commitment problem in Eurasian energy transit. *Review of International Political Economy* .

Syroezhkin, K. (2001). Kazakhstan's security policy in the Caspian Sea region. In G. Chufrin, *The Security of the Caspian Sea Region* (pp. 212-230). New York: Oxford University Press.

TAP AG. (2016). *About us.* Retrieved January 3, 2016, from Trans Adriatic Pipeline: http://www.tap-ag.com/about-us

TAP AG. (2013, July 30). *BP, SOCAR, Total and Fluxys join the TAP project.* Retrieved August 1, 2013, from TAP:

http://www.tap-ag.com/news-and-events/2013/07/30/bp-socar-total-and-fluxys-join-the-tap-project

TAP AG. (2016, January 5). *The Big Picture.* Retrieved January 5, 2016, from TAP-AG: http://www.tap-ag.com/the-pipeline/the-big-picture

TAP AG. (2016). *Trans Adriatic Pipeline route.* Retrieved January 5, 2015, from Trans Adriatic Pipeline: http://www.tap-ag.com/the-pipeline/route-map

Tarr, D. (2007). Russian Accession to the WTO: An Assessment. *Eurasian Geography and Economics, 48* (3), 306-319.

Trenin, D. (2009). Russia's Sphere of Interest, Not Influence. *Washington Quarterly , 32* (4), 3–22.

Tsereteli, M. (2011). Connecting Caspian Gas to Europe: No large scale infrastructure development in near future. *Turkish Policy Quarterly , 9* (2).

Tsygankov, A. P. (2001). *Pathways after Empire: National Identity and Foreign Economic Policy in the Post-Soviet World.* Lanham, MD: Rowman and Littlefield.

UNDP. (2004). *World energy assessment: overview.* United Nations Development Program . New York: UNDP.

Van der Linde, C., Amineh, M., Correlje, A., & Jong, D. (2004). *Study on Energy Supply Security and Geopolitics.* Institute for International Relations "Clingendael", Clingendael International Energy Programme. The Hague: CIEP.

Van Der Meulen, E. (2009). Gas Supply and EU–Russia Relations. *Europe-Asia Studies, 61* (5).

Van Langenhove, L. (2011). *Building Regions: The Regionalization of the World Order.* Farnham: Ashgate Publishing Ltd.

Vatansever, A. (2010). *Russia's Oil Exports: Economic Rationale versus Strategic Gains.* Washington DC: Carnegie Endowment International Peace Paper.

Vernon, R. (1971). *Sovereignity at Bay.* New York.

Verrastro, F., Ladislaw, S., Frank, M., & Hyland, L. (2010). *Geopolitcs of Energy: Emerging Trends, Changing Landscapes, Uncertain Times.* Center for Strategic and International Studies (CSIS), Energy and National Security Program. Washington DC: CSIS.

Victor, D., Jaffe, A., & Hayes, M. (2006). *Natural Gas and Geopolitics: From 1970 to 2040.* New York: Cambridge University Press.

Viëtor, M. (2011). *Energiesicherheit für Europa: Kernenergie und Erdgas als Brücken-technologien.* Baden-Baden: Nomos.

Volovik, Y. (2011). *Overview of Regional Transboundary Water Agreements, Institutions and Relevant Legal/Policy Activities in Central Asia .* Almaty: UNDP.

Vzgliad. (2014, June 27). *Chizhov: there is no proof that South Stream is not compliant with the EU law '*, Retrieved 2015, from www.vz.ru/news/2014/6/27/693041.html.

Walde, T. (2008). Renegotiating Acquired Rights in the Oil and Gas Industries: Industry and Political Cycles Meet the Rule of Law. *Journal of World Energy Law & Business, 1* (1).

Wallerstein, I. (1989). *The Modern World-System, III: The Second Era of Great Expansion of the Capitalist World Economy, 1730s – 1840s.* New York: Academic Press.

Walsh, P. (2013). Norbert Elias and Hannah Arendt on Philosophy, Sociology and Sience. In F. Dépelteau, & T. S. Landini, *Norbert Elias and Social Theory.* NY: Macmillan Palgrave.

Waltz, K. (1979). *Theory of International Politics.* Reading, M.A., USA: ddison-Wesley Publishing Company.

Weisbrode, K. (2001). *Central Eurasia: Prize or Quicksand?.* Oxford: The International Institute for Security Studies/Oxford University Press.

Williamson, O. (1975). *Markets and Hierarchies.* New York: The Free Press.

Winrow, G. (2004). Turkey and the East-West Gas. *Turkish Studies , 5* (2), 23-42.

Woodhouse, E. J. (2006). The Obsolescing Bargain Redux: Foreign Investment in the Electrical Power Sector in Developing Countries. *International Law and Politics , 38* (121), 121–33.

Yafimava, K. (2013). *The EU Third Package for Gas and the Gas Target Model: major contentious issues inside and outside the EU.* Oxford: OIES.

Yergin, D. (2006). Ensuring Energy Security. *Foreign Affairs , 85* (2), 69-82.

Yergin, D. (2008). *The Prize: The Epic Quest for Oil, Money, and Power* (Reissue edition ed.). New York: Free Press.

Yergin, D. (2011). *The Quest: Energy, Security, and the Remaking of the Modern World.* New York: Penguin Books.

Young, R. (2009). *Energy Security: Europe's New Foreign Policy Challenge.* London: Routledge.

Zhiltsov, S. (2014). The Caspian Region at the corssroad of geopolitical startegies. *Cnertal Asia and the Caucasus, 15 (1)*